THE ELECTRIC SELF

How Your Brain's Hidden Electrical Universe Creates Everything You Think, Feel, Remember, and Are

By

Mohammad Umar

For permission requests, speaking engagements, and all other correspondence, please contact the author directly.

First Edition, 2026
Printed in India
Also available in digital editions worldwide

AF720549

PREFACE

The Most Personal Science

THERE IS A FACT ABOUT you that I want to tell you before we go any further. It is not a comfortable fact. It is not the kind of thing you are likely to have heard stated plainly, without hedging or softening, in any classroom or doctor's office or popular article about the brain. But it is true, and it matters, and I believe that sitting with its strangeness is the most important first step you can take toward understanding everything that follows in this book.

The fact is this. Every memory you have ever made exists inside your skull as a pattern of electrical and chemical changes in biological tissue. Every person you have ever loved is represented in your brain as a configuration of synaptic weights, a particular arrangement of connection strengths between neurons that encodes something of what that person meant to you, who they were, how they made you feel. Every dream you have ever had was nothing more and nothing less than your brain generating electrical signals in the absence of sensory input, weaving those signals into narrative and imagery and emotion and then, when you woke, mostly dissolving them again like a sand mandala swept into the sea. Every moment of grief you have endured, every burst of joy, every flash of creative inspiration, every experience of beauty or terror or transcendence, each of these was, at the level of physics, a pattern of ions moving across membranes, of electrical potentials rising and falling, of neurotransmitters crossing synaptic gaps, of action potentials propagating along axons at speeds of up to 120 metres per second. There is nothing else. There is no additional substance. There is no ghost in the machine, no soul-stuff animating the biological hardware. There is the hardware. And then there is, somehow, astonishingly, irreducibly, you.

I want to be honest with you about something. When I first truly understood this, not just as an abstract proposition but as a biological reality that I could trace through the mechanisms and the measurements and the experiments, I found it deeply unsettling. Not because it diminished my sense of my own experience, though I feared it might. But because it raised a question so large and so strange that I could not look away from it, and once I could not look away from it, I found that it was in fact the most interesting question I had ever encountered in all my years of reading about science. The question is this: how does it happen? How does electricity become experience? How do the physical mechanisms of the firing neuron become the felt quality of a summer afternoon, or the specific ache of missing someone, or the way a piece of music can make you feel as though you are remembering something you have never actually experienced? How does the biological become the personal? How does matter become mind?

Nobody knows. That is the honest answer, and I will not dress it up. The question has a formal name in philosophy of mind. The philosopher David Chalmers, whose 1995 paper on the subject remains one of the most cited in all of academic philosophy, called it the Hard Problem of consciousness, and the name has stuck because it captures something important: there are many questions about the brain that are difficult but tractable, problems of mechanism and architecture and function that neuroscience is steadily answering. And then there is this one, which is different in kind, which resists the methods that work so well for everything else, which continues to defy resolution despite more than a century of increasingly sophisticated neuroscience and millennia of philosophical reflection before that.

But here is what I also want to tell you, what drove me to spend years researching and writing this book: the search for the answer has already produced discoveries so extraordinary, so beautiful, and so immediately consequential for human health and human freedom, that the journey is worth taking even without a guaranteed destination. In the course of trying to understand how electricity becomes experience, neuroscience has mapped the neural architecture of memory and emotion with stunning precision. It has identified the specific brain circuits that malfunction in depression, anxiety, epilepsy, Alzheimer's disease, Parkinson's disease, and schizophrenia. It has developed electrical and magnetic therapies that can heal those malfunctioning circuits in ways that decades of pharmacology could not. It has created devices that allow paralysed patients to control computers, robotic arms, and their own voices using nothing but their thoughts. It has developed systems that can reconstruct the words a person is thinking from the electrical activity of their brain alone. It has revealed that psychedelic compounds heal treatment-resistant depression not by adding new chemistry but by temporarily dissolving the brain's rigid electrical habits and allowing new patterns to form. And it is now on the verge of a technology that will merge human and machine intelligence in ways that will raise questions about identity, freedom, and what it means to be a person that no previous generation of human beings has ever had to confront.

This book is about all of that. But it is also, at every point, about something more personal than any technology or therapy or scientific finding. It is about you. About what you are made of, at the deepest level that science has yet reached. About why your memories work the way they do and sometimes fail the way they do. About why your emotions can feel so overwhelming and yet so beyond your conscious control. About why sleep matters more than almost anything else you can do for your brain. About what is actually happening in the brains of people who suffer from depression or anxiety or dementia, and why it matters enormously that we understand these conditions as disorders of biological machinery rather than failures of character or will. And about what the coming age of neural technology means for the future of your most intimate possession: the private electric universe inside your skull.

I should tell you something about how this book is written, because I have made a deliberate choice that I want to explain. Each chapter is written in a different style and a different voice, suited to the particular material it covers. The historical chapters are novelistic and character-driven, because history is nothing if not the story of particular people doing particular things in particular moments that changed everything. The science chapters are precise and metaphor-rich, because the best science writing I have ever read treats the reader as an intelligent adult who does not need the complexity removed but does need it illuminated. The chapters on brain disorders are compassionate and human, centering the experiences of patients, because the science of neurological disease is meaningless if it is not connected to the suffering that motivates it and the hope that it generates. And the final chapters are frankly philosophical and speculative, because the future of the electric brain is arriving so fast that speculation is not an indulgence but a necessity.

There will be moments in this book where the science is genuinely difficult, where the mechanisms are intricate and the terminology unavoidable. I have put a glossary at the back of the book, and I encourage you to use it freely, without embarrassment. There will also be moments where the science shades into territory that is uncertain, contested, or preliminary, and in those moments, I will tell you so clearly. I have no interest in presenting the current state of neuroscience as more complete or more certain than it is. The brain is the most complex object in the known universe, and the science of understanding it is still, in many of its most important domains, in its infancy. The honest acknowledgement of that fact is not a weakness. It is the only intellectually respectable position available.

One last thing, and then we begin. I want to ask you to do something as you read this book. I want to ask you to periodically put it down, not permanently, not in frustration, but deliberately, and to simply be aware of what is happening in your own mind as you do so. Notice the sensation of thinking. Notice the way attention moves, shifts, narrows, expands. Notice the emotional quality of different ideas as they arrive, the way some thoughts feel warm and others feel cold, the way memory intrudes unbidden into the stream of present experience. Notice the you that is noticing all of this. And then, when you pick the book up again, consider that everything you just experienced, the whole rich, intimate, irreducibly first-person texture of your conscious life, was and is and will always be electricity. Patterned, biological, ancient electricity, refined by hundreds of millions of years of evolution, operating at the intersection of chemistry and physics and information and life. And that this fact, far from reducing you to something small or mechanical, places you at the most astonishing frontier in all of science: the edge where the physical world looks at itself and wonders what it is.

Mohammad Umar

February 2026

INTRODUCTION

The Universe Looking at Itself

SOMEWHERE IN YOUR SKULL, right now, a storm is happening. Not a metaphorical storm. Not a poetic approximation of something that is really going on at another level of description. A genuine electrical storm: approximately 86 billion neurons firing action potentials at rates of up to several hundred times per second, each firing event a precisely timed voltage spike lasting about one millisecond, each spike propagating along an axon at speeds of up to 120 metres per second, each arrival at a synaptic terminal triggering the release of neurotransmitter molecules across a twenty-nanometre gap to the next neuron in the chain. There are, in your brain right now, approximately 100 trillion of these synaptic connections, each one capable of being strengthened or weakened by the pattern of electrical activity that passes through it, each one thereby constituting a kind of memory of every experience the brain has processed since the synapses first formed.

This electrical storm is you. Not a part of you, not the mechanism underneath you, not the substrate of something else that is really you. This storm, in its full complexity and its precise moment-by-moment configuration, is identical with everything that you are: your memories of your childhood and the faces of the people you love, your capacity for language and mathematics and music, your political opinions and your aesthetic preferences and the things that make you laugh, your fears and your desires and the specific quality of your character that makes you recognizable to the people who know you as yourself rather than anyone else. There is nothing additional. There is the electrical storm in the brain. And then there is, somehow, astonishingly, irreducibly, you. The storm is the self. This book is the attempt to understand how.

What This Book Is

The Electric Self is a popular science book about the neuroscience of brain electricity. That description is accurate but insufficient. It is also a book about consciousness and the deepest mystery in all of science. It is a book about memory and identity and what it means that the self is a biological construction rather than a metaphysical given. It is a book about the extraordinary range of neurological and psychiatric conditions that result when the brain's electrical machinery breaks down, and about the equally extraordinary range of electrical, chemical, and technological interventions that medicine and neuroscience have developed to repair it. And it is, in its final section, a book about the future: about the brain-computer interfaces that are already allowing paralyzed patients to type with their thoughts and speak with their minds, about the neural decoding technology that can reconstruct the semantic content of what a person is thinking from their brain activity alone, and about the governance questions this technology is raising that no previous generation of human beings has had to confront.

The book is organized into five parts, each corresponding to a distinct domain of the science. Part One, The Spark of Existence, tells the history of bioelectricity: the 244-year arc from Luigi Galvani's discovery that electricity animates living tissue to the Hodgkin-Huxley equations that gave science its first complete mathematical description of the action potential, the fundamental unit of neural computation. Part Two, The Electrical Architecture of the Mind, examines how electricity creates memory, emotion, sleep, and consciousness, exploring each of these domains through the lens of the neuroscience that has illuminated them and the deep questions that remain unanswered. Part Three, When the Electricity Goes Wrong, reframes the major neurological and psychiatric conditions, epilepsy, depression, anxiety, Alzheimer's disease, and Parkinson's disease, as disorders of the brain's electrical machinery rather than failures of character or will. Part Four, Rewiring the Brain, covers the technologies of electrical and chemical intervention: deep brain stimulation and transcranial magnetic stimulation, brain-computer interfaces and neural decoding, and the psychedelic medicines whose mechanism of action is now being understood in terms of what they do to the brain's electrical organization. And Part Five, The Electric Future, examines the coming age of neural augmentation and asks the governance questions that this age is making urgent: who owns neural data, who has the right to alter the electric self, and what protections human beings will need in a world where the most intimate facts about their inner lives are accessible to anyone with the right technology and the right access.

Each chapter is written in a different style and voice; chosen to suit the particular material it covers. The historical chapters of Part One are novelistic and character-driven, centered on the specific scientists whose discoveries shaped the field. The science explanations of Part Two are precise and metaphor-rich, written for the intelligent general reader who wants to understand the mechanisms without needing specialized training to follow the argument. The clinical chapters of Part Three are compassionate and human, centering the experiences of patients alongside the science of the conditions they live with. The technology chapters of Part Four are written with the urgency and pace of science journalism, because the developments they describe are happening now and their implications are immediate. And the final chapters of Part Five are frankly philosophical and essayistic, because the future they describe requires not just information but reflection.

Science tells us what is possible. Philosophy, politics, and ethics determine what of the possible we choose to do.

Why Now

The science of the electric brain has been developing for 244 years, but it has never moved as fast as it is moving now, and the convergence of several simultaneous revolutions is making the current moment particularly consequential for anyone who wants to understand where neuroscience is taking us.

The first revolution is technological. The tools available for studying the brain, and for intervening in it, have improved more in the past twenty years than in the previous two centuries combined. Functional MRI now allows researchers to measure the activity of cubic-millimeter volumes of brain tissue in real time across the whole brain simultaneously. High-density electrocorticography, in which electrode arrays are placed directly on the brain's surface, records neural activity with millisecond temporal resolution and millimeter spatial resolution across large cortical areas. Calcium imaging allows researchers to track the activity of individual identified neurons in living animals across weeks and months of behavior. Single-cell RNA sequencing has revealed the full molecular diversity of the brain's cell types in unprecedented detail. And implanted brain-computer interfaces are now recording from thousands of neurons simultaneously in human patients, providing data of a quality and a richness that makes the earliest BCI experiments look, by comparison, like trying to understand an orchestra by listening to it through a wall.

The second revolution is computational. The machine learning algorithms that are transforming every other field of science and technology have arrived in neuroscience with particular force, because the brain's data is exactly the kind of high-dimensional, complex, temporally structured data that deep learning methods are designed to analyze. Neural decoding algorithms can now extract semantic content from fMRI recordings, reconstruct visual imagery from brain activity, and decode intended speech from cortical signals with accuracy that was unthinkable a decade ago. The combination of better brain measurement and better data analysis is producing discoveries at a rate that makes the field almost impossible to summarize without immediate obsolescence.

The third revolution is clinical. The first approved brain-computer interface devices are now in human patients. The first anti-amyloid antibody treatments for Alzheimer's disease have demonstrated, for the first time in the disease's history, that the progression of dementia can be measurably slowed by targeting its biological mechanisms. Psilocybin has received FDA Breakthrough Therapy designation for treatment-resistant depression and is in Phase 3 clinical trials. MDMA has demonstrated sixty-seven percent PTSD remission rates in Phase 3 trials. Deep brain stimulation is being actively investigated for conditions ranging from depression and OCD to anorexia nervosa and chronic pain. Transcranial magnetic stimulation is approved for an expanding list of indications and is increasingly accessible as a front-line rather than last-resort intervention. The neuroscience of the past decade is beginning to produce a clinical dividend that the neuroscience of the previous century largely promised but rarely delivered.

Together, these three revolutions are producing a moment of genuine transformation, one in which the understanding of the electric brain is advancing faster than our ability to fully process its implications. This book is an attempt to help with that processing: to make the science accessible, to make the history vivid, to make the clinical realities humanly present rather than abstractly statistical, and to make the governance questions concrete enough to be acted upon rather than merely worried about.

A Note on Certainty and Uncertainty

One of the most important commitments this book makes is to honesty about what is known and what is not. Neuroscience is a young science. For all the extraordinary progress of the past century, there is vastly more that is unknown than known about how the brain works, and the domains in which the unknowns are largest are, characteristically, the domains that matter most. We do not know how consciousness arises from neural activity. We do not have a complete understanding of the mechanisms of any major psychiatric or neurological disorder. We do not have effective treatments for most neurodegenerative diseases. We do not know whether any current theory of consciousness is correct. We do not know, with confidence, what the long-term consequences of implanting electronic devices in the human brain will be for the individuals who receive them.

This honest acknowledgement of uncertainty is not a weakness of the book or of the science. It is the only intellectually respectable position available. Science advances by acknowledging what it does not know and systematically working to find out. The premature certainty that popular science sometimes produces, the confident assertion that a single study has resolved a question that is in fact still deeply contested, does a disservice to both the science and the reader. Where the evidence is strong and the consensus is clear, this book says so. Where the evidence is preliminary, or the interpretation contested, or the question genuinely open, it says that too.

There is one domain in which the uncertainty is particularly important to acknowledge: the clinical chapters of Part Three describe conditions, depression, anxiety, Alzheimer's disease, epilepsy, that affect hundreds of millions of people and that carry enormous stigma, suffering, and social consequence. These chapters approach those conditions from a neuroscientific perspective that frames them as disorders of the brain's biological machinery. This framing is, in this author's view, both scientifically accurate and ethically necessary: it is the basis for reducing stigma, for developing better treatments, and for treating people with these conditions with the respect and the compassion that their suffering deserves. But framing a condition as biological does not mean reducing the person who has it to their biology, or suggesting that the lived experience of the condition is fully captured by any description of neurons and circuits and oscillations. The science and the humanity are not in competition. They are both required, and this book attempts to hold both simultaneously throughout.

How to Read This Book

The Electric Self is written to be read from beginning to end, as the chapters build on each other and the later sections draw on concepts established in the earlier ones. But each chapter is also designed to be substantially self-contained, and a reader who is primarily interested in a particular topic, memory, or sleep, or brain-computer interfaces, or psychedelics, can begin with that chapter and follow the cross-references to related material elsewhere in the book.

A glossary of key technical terms appears at the back of Volume Three, organized alphabetically. I encourage you to use it freely and without embarrassment: the terminology of neuroscience is dense and specific, and looking up a term in the glossary is not a sign of inadequate background but a sign of the kind of careful reading that the science deserves.

Each chapter opens with a specific human being: a scientist or a patient or a research subject who is connected to the material the chapter covers. This is a deliberate structural choice. Neuroscience is sometimes written in a way that treats the brain as an object of study divorced from the people who have brains and whose lives are shaped by the science. This book takes the opposite approach: the science is always, at some level, about someone. Galvani's observations changed how we understand what life is. Henry Molaison's tragedy illuminated the neural architecture of memory. Helen's deep brain stimulation surgery raised questions about identity that philosophy has not yet answered. Ann's restored speech after eighteen years of silence is both a technological achievement and a human story of loss and recovery.

The science is richer and truer when it is connected to the human beings whose lives it has shaped and to whom its continued advance is dedicated.

Finally, I want to say something about what I hope this book produces in the reader, beyond information. I hope it produces a particular quality of attention to your own experience. The fact that everything you think and feel and remember is electrical does not make those thoughts and feelings and memories less real or less meaningful. It makes them more astonishing: it means that the universe has produced, through four billion years of biological evolution, an entity capable of sitting in a chair and wondering what it is. You are the universe doing that, right now. The mechanisms that make this possible are described in the pages that follow. The mystery of how those mechanisms produce the experience of being you is still, honestly, open. I find that open mystery to be the most beautiful fact about the current state of human knowledge. I hope, by the time you reach the final pages of this book, you will find it beautiful too.

Mohammad Umar

February 2026

PART ONE

THE SPARK OF EXISTENCE

The Discovery of Bioelectricity and the Birth of a Science

Part Introduction

A Jolt That Changed Everything

The autumn of 1780 was not a particularly remarkable season in Bologna. The city moved through its customary rhythms: the university opened its doors to another cohort of students, the markets filled and emptied with the week, the air cooled toward evening in the way that Italian autumn evenings do, smelling of stone and wood smoke and the particular dustiness of old libraries. In a house near the university, a professor of anatomy named Luigi Galvani was working late, as he usually did, surrounded by the equipment of his investigations: copper wire, brass hooks, glass vials, and the dissected remains of frogs, which were his preferred experimental subject because their large sciatic nerve and the powerful muscles of their hind legs made visible even the smallest electrical effects.

What happened next, on an evening that Galvani himself could not have imagined would matter to anyone beyond his immediate circle of colleagues, was one of those accidents that science periodically requires to break through the walls of its own assumptions. An assistant, whose name history has not preserved, was working at a nearby table. He touched a metal scalpel to the exposed sciatic nerve of a dead frog. At the same moment, elsewhere in the room, another member of Galvani's household drew a spark from an electrical machine, a device used to generate static electricity for various experiments. The frog's leg kicked. It was dead. It had been dead for hours. And it kicked.

Galvani stared. His wife, Lucia Galeazzi Galvani, who was present in the room and who was herself a trained scientist and an important intellectual partner in her husband's work, is said to have gasped. The assistant stepped back from the table. The leg was still. The room was quiet. And then Galvani did what every great scientist does in such a moment: he did not dismiss what he had seen, did not attribute it to error or coincidence or the flickering of tired eyes. He investigated it.

He repeated it. He varied the conditions. He kept meticulous records. And over the years that followed, he gradually assembled the body of observation and experiment that would, in 1791, become one of the most consequential publications in the history of science.

What Galvani had discovered, or believed he had discovered, was what he called animal electricity: the idea that living tissue itself generates and uses electrical forces, and that this electrical quality of living matter is somehow essential to the phenomenon of life. He was not entirely right about the mechanisms, and the next two decades would see his ideas vigorously contested in one of the great scientific debates of the age. But the core intuition was correct, and it was revolutionary, and it opened a door through which the entire science of neuroscience would eventually pass.

Part One of this book tells the story of that passage: the long, strange, brilliant, sometimes deeply misguided history of humanity's attempt to understand the electricity of life. It moves from Galvani's Bologna to the great physiological laboratories of nineteenth-century Germany, where the mechanisms of the nerve impulse were first measured and mathematically described. It follows the eccentric genius of Hans Berger, who listened to the living human brain for the first time in 1924 and spent years being laughed at for what he heard. And it arrives at the twentieth century's greatest triumph in neuroscience: the work of Alan Hodgkin and Andrew Huxley, who unlocked the molecular mechanism of the action potential and gave science, for the first time, a complete mathematical description of the fundamental unit of all neural computation.

Throughout this historical narrative, a central theme will emerge, and it is the same theme that runs through every part of this book: the tension between what we can measure and what we cannot fully explain. Science has been extraordinarily successful at describing the mechanisms of the brain's electricity, the ions, the channels, the potentials, the frequencies, the networks. What it has not yet explained, and what makes the brain's electricity unlike any other electrical phenomenon in the known universe, is why any of it feels like anything at all. That tension between the measurable and the experienced is not a gap in the science that future research will tidy up. It is the hard edge of the deepest mystery in all of human knowledge, and learning to sit with it, to hold the mechanisms and the mystery together without collapsing one into the other, is part of what this book asks of its reader.

CHAPTER ONE

Dead Frogs and Living Electricity

The Discovery of Bioelectricity and the Scientists Who Were Laughed At

THE FROG HAD BEEN DEAD for most of the morning. Galvani had prepared it himself, with the careful hands of a man who had dissected hundreds of animals over two decades of anatomical research, exposing the sciatic nerve along the length of the thigh and securing the specimen to the wooden board with small brass pins. It lay there in the October light, inert and irreversible, as dead as the stone of the table it rested upon. And then it moved.

That single involuntary convulsion of a dead frog's leg is, in retrospect, one of the hinge moments of Western science. Not because Galvani was the first person to notice that electricity could produce muscular contraction in dead tissue; that observation had been made earlier by the English physician William Cadogan, among others. But because Galvani, unlike his predecessors, refused to treat the observation as a curiosity and instead embedded it in a program of rigorous, systematic investigation that lasted more than a decade and produced findings that permanently altered the human understanding of what life is and how it works.

Luigi Galvani was born in Bologna in 1737, the son of a goldsmith who was prosperous enough to send his son to the university in that city, which was one of the oldest and most prestigious seats of learning in Europe. Galvani trained first in medicine and then in anatomy, and he rose to become a professor at the university while also practicing as a physician. He was, by all accounts, a man of great personal warmth and intellectual humility, qualities that would serve him well in the decade of controversy that his discoveries produced, and that would also, in the end, be insufficient to protect him from the consequences of being on the wrong side of a dispute with a more famous and more politically astute opponent.

The Accidental Beginning

The formal beginning of what would become the science of bioelectricity is usually dated to the autumn of 1780, though Galvani's notebooks suggest that he had been observing and recording electrical effects in animal tissue for some time before the famous evening of the kicked frog. What made that evening significant was not the observation itself but the conjunction of circumstances that made its meaning unmistakable: the electrical machine running in the same room as the frog, the metal instrument connecting the nerve, the spark and the kick occurring simultaneously in a way that could not be attributed to coincidence.

Galvani's subsequent experiments were designed with elegant simplicity. He wanted to know whether the electricity that caused the muscle to contract came from outside the animal, as the electrical machine hypothesis would suggest, or from inside it, as his developing theory of animal electricity proposed. He devised increasingly clever experimental setups to distinguish between these possibilities. He hung frog legs on brass hooks from an iron railing in his garden, and observed that they twitched during thunderstorms and sometimes even in dry weather when the atmospheric electricity was elevated, a result that seemed to support the idea of external electrical influence. But then he noticed something that the external-electricity hypothesis could not easily explain: the frog legs also twitched when he brought the nerve into contact with the muscle through a conducting arc, even in the complete absence of any external electrical source. The animal was generating its own electricity. The electricity was coming from within.

He published his findings in 1791 in a paper with the magnificently baroque Latin title Commentary on the Effect of Electricity on Muscular Motion. The paper caused an immediate sensation in the scientific world of late eighteenth-century Europe, and not only because of its findings. Galvani wrote with unusual care and honesty about the process of his investigation, including his mistakes and revisions and the alternative hypotheses he had considered and rejected. This was not how scientific papers were typically written in the eighteenth century, and it gave the Commentary a vividness and a credibility that purely factual reports often lacked.

The Man Who Argued Back

The response that mattered most, and that would define Galvani's legacy for a generation, came not from a supporter but from a critic: Alessandro Volta, professor of experimental physics at the University of Pavia and the most celebrated natural philosopher in Italy. Volta was Galvani's intellectual peer and in some respects his superior: he was a more rigorous experimentalist, a more polished writer, and far more connected to the networks of European scientific power that determined whose ideas received attention and whose were ignored. He was also, as it turned out, partly right about what Galvani had observed, in a way that made Galvani's interpretation partly wrong, though the full picture would take more than a century to emerge.

Volta's objection to Galvani's theory of animal electricity was technical but fundamental. The twitching of the frog's leg, Volta argued, was not caused by electricity generated by the animal itself. It was caused by the contact between two different metals, the brass hook and the iron railing, which in the presence of a moist conductor such as a frog's nerve and muscle created what we would now recognize as a simple electrical cell, a primitive battery. The animal was not the source of the electricity. It was the detector of electricity generated by purely physical and chemical means. Galvani's frog legs were, in Volta's framing, exquisitely sensitive electrical instruments, not electrical generators.

The debate that followed between Galvani and Volta was one of the most intellectually consequential in the history of science, and it was thoroughly public: conducted in papers, in letters, in demonstrations before academies of science across Europe, with both men performing new experiments designed to defeat the other's position. The scientific community was largely won over by Volta, whose prestige and rhetorical skill gave him significant advantages. And Volta was spectacularly vindicated when in 1800 he constructed a column of alternating zinc and copper discs separated by cloth soaked in salt water and produced a continuous electrical current: the voltaic pile, the world's first electric battery. The invention was received as a triumph, and it was. But it was also, paradoxically, a demonstration that Galvani had been essentially correct about the existence of animal electricity, since the voltaic pile worked on exactly the same electrochemical principles that operate in living nerve and muscle tissue.

Galvani himself did not live to see his vindication. He died in 1798, in obscurity and poverty, his academic positions stripped from him after he refused, on grounds of conscience, to swear allegiance to the Cisalpine Republic imposed on Bologna by Napoleon's armies. He died a few years before the publication that might have restored his reputation. History remembers both men, but for different reasons: Volta for the battery, which launched the age of electrical technology; Galvani for the observation that started everything, which is commemorated in the English language every time someone speaks of being galvanized into action.

The Romantics and the Resurrection of the Dead

The debate between Galvani and Volta produced a cultural aftershock that reverberated far beyond the walls of scientific academies. The idea that electricity might be the animating force of life, that the difference between a living body and a dead one might be the presence or absence of this invisible, crackling, mysterious force, seized the European imagination at exactly the moment when Romanticism was teaching that imagination to reach for the most extreme and transgressive possibilities of human experience.

Giovanni Aldini, Galvani's nephew and the most theatrical champion of his uncle's ideas, conducted a series of public demonstrations in the first decade of the nineteenth century that drew enormous crowds and even more enormous controversy. In London in 1803, Aldini obtained the body of a recently hanged murderer named George Forster and connected it to a voltaic pile in front of an audience that included members of the Royal College of Surgeons, several members of Parliament, and a large number of journalists. When he applied electrical contacts to the dead man's face, the muscles of the jaw contracted and the lips moved. When he applied current to the ear and the rectum simultaneously, the body arched as if in agony. When he stimulated the phrenic nerve, the chest heaved as if breathing. The Times of London reported the next day that several members of the audience had fainted and that one had fled the room in terror, convinced that the corpse was about to sit up and walk.

It did not. But the demonstration planted a seed that would flower, fifteen years later, in one of the most enduring works of the Romantic imagination. Mary Shelley was nineteen years old when she began writing Frankenstein, and the conversations she had heard the previous summer at the Villa Diodati near Lake Geneva, where the poet Lord Byron held court and where the fashionable scientific ideas of the day were discussed with equal parts enthusiasm and horror, included animated discussion of Galvani and Aldini and the possibility of electrically reanimating dead tissue. Her novel, published in 1818 under the full title Frankenstein, or The Modern Prometheus, is many things: a meditation on the responsibilities of scientific creation, a story about the relationship between creator and creation, a gothic thriller of enduring power. But it is also, unmistakably, a response to the science of bioelectricity, and its influence on how subsequent generations have thought about the relationship between electricity and life has been immeasurable.

The German Giants

The next great chapter in the story of bioelectricity was written not in the drawing rooms of Romantic poets but in the physiological laboratories of nineteenth-century Germany, where a generation of scientists armed with increasingly sensitive instruments began the slow, rigorous, often maddeningly technical work of understanding exactly how and why living tissue generates electricity and what that electricity does.

The central figure of this chapter is Emil du Bois-Reymond, a man whose name is less well known than it deserves to be, given that he established the electrical theory of the nerve impulse on a foundation of experimental evidence so solid that it has never seriously been challenged. Du Bois-Reymond was born in Berlin in 1818 to a Swiss father and a Huguenot mother, and he spent his entire scientific career at the University of Berlin, where he became one of the most influential physiologists of the nineteenth century. He was also, by all accounts, a man of formidable personality, opinions, and oratorical gifts, who could fill a lecture hall and empty an opposing argument with equal facility.

What du Bois-Reymond contributed to science was a combination of technological innovation and meticulous experimental rigor that transformed the study of bioelectricity from a collection of suggestive observations into something approaching a quantitative science. He built galvanometers of unprecedented sensitivity, devices that could detect electrical currents far too small to produce any visible physical effect, and used them to demonstrate beyond reasonable doubt that resting nerve and muscle tissue maintains a steady electrical potential across its membrane, and that when a nerve is stimulated or a muscle contracts, this potential changes in a specific and reproducible way. He called this change the negative variation, and it is what we now call the action potential, the fundamental unit of neural signalling. Du Bois-Reymond could not explain the mechanism of the action potential, because the concept of ions and cell membranes did not yet exist in biology. But he established its existence with extraordinary precision, and his 1848 masterwork Investigations into Animal Electricity remains one of the great monuments of nineteenth-century science.

A decade later, his friend and sometime rival Hermann von Helmholtz made the measurement that completed the foundation: the speed of the nerve impulse. Helmholtz was, by any measure, one of the most extraordinarily gifted scientists of the nineteenth century, making fundamental contributions to thermodynamics, optics, acoustics, and the physiology of perception. His measurement of nerve conduction velocity, published in 1850, was almost embarrassingly simple in its experimental design: he stimulated a frog's sciatic nerve at different distances from the muscle it controlled and measured the time between the stimulus and the muscle contraction. By comparing the times for stimulation at different points along the nerve, he could calculate how fast the impulse was traveling. The answer was approximately 27 metres per second, a figure that astonished the scientific world.

The astonishment was not, as one might expect, because the speed was so great. It was because it was so small. The prevailing assumption, based on analogy with the propagation of electrical signals in metal wires, where the speed approaches that of light, was that the nerve impulse must be effectively instantaneous. Helmholtz's measurement showed that it was nothing of the kind: it was slower than a horse at full gallop, slower than sound in water, slower than a rifle bullet. This meant that the nerve impulse was not a simple electrical conduction of the kind that occurs in a wire. It was something biological, something fundamentally different, something that used electricity but was not reducible to it. The nerve impulse was a wave of electrochemical change that propagated itself along the nerve by a process of self-renewal, using the chemistry of the living membrane to regenerate itself at every point along its path. Helmholtz could not explain this mechanism either. That explanation would have to wait for the twentieth century, for the right animal in the right laboratory, and for two of the most remarkable scientists who ever lived.

The Squid That Saved Neuroscience

The Atlantic longfin squid, known by its scientific name Doryteuthis pealeii and by fishermen along the northeastern coast of the United States as the brief squid, is not, by most aesthetic standards, a prepossessing animal. It is perhaps 30 centimetres long, torpedo-shaped, pale and translucent when alive, capable of remarkable camouflage and of squirting ink when alarmed, and it is caught in enormous quantities off the Massachusetts coast where it forms an important part of the local fishing economy. It is also, as it turned out, the animal that made possible one of the greatest achievements in the history of neuroscience, because it possesses something that no vertebrate animal has: a giant axon.

The giant axon of the squid is a nerve fiber so large, by the standards of the animal kingdom, that it is visible to the naked eye. While a typical mammalian nerve fiber might be a few micrometres in diameter, the squid's giant axon can reach a millimetre in width: enough to insert an electrode into its interior and measure the electrical potential across the membrane directly, from inside the cell, rather than inferring it from recordings taken at a distance. This technical advantage was the key that unlocked the mechanism of the action potential.

The scientists who used this key were Alan Hodgkin and Andrew Huxley, working at the Marine Biological Laboratory in Plymouth, England, and at the Woods Hole Oceanographic Institution in Massachusetts, during the years immediately before the Second World

War interrupted their work and in the years immediately after it. Hodgkin and Huxley were in many ways an unlikely pair: Hodgkin was the more intuitive and experimentally gifted, Huxley the more mathematically rigorous and analytically powerful. Together they were extraordinary. Between 1939 and 1952, they developed a series of increasingly refined experimental techniques and mathematical models that eventually produced a complete quantitative description of the action potential: how it is initiated, how it propagates, what ions are moving across the membrane at each stage of the process, and what electrical and chemical forces drive those movements.

The Hodgkin-Huxley model, published in a series of five papers in the Journal of Physiology in 1952, is widely regarded as one of the most important contributions to biology in the twentieth century. It showed that the action potential is produced by the sequential opening and closing of specific protein channels in the cell membrane, which allow first sodium ions to rush into the cell, driven by both electrical and concentration gradients, and then potassium ions to rush out, restoring the original electrical state. The whole process takes about a millisecond. It consumes a tiny amount of energy. And it can be repeated essentially without limit, hundreds of times per second, as long as the cell remains alive and the sodium-potassium pump continues to maintain the ionic gradients that make each action potential possible. For this work, Hodgkin and Huxley received the Nobel Prize in Physiology or Medicine in 1963, an honour that no one in the field regarded as premature or excessive.

The Meaning of the Spark

The discovery of the action potential mechanism did not solve the problem of consciousness. It did not explain how the firing of a neuron contributes to the experience of a thought or a feeling. It did not bridge the gap between the physical and the personal, between the ion channel and the inner life. But it did something that was absolutely necessary before those harder questions could even be properly asked:

it established, beyond any remaining reasonable doubt, that the language of the nervous system is electricity, and that this electrical language follows knowable physical laws.

Everything that has been discovered about the brain in the seventy years since Hodgkin and Huxley published their equations has confirmed and extended this fundamental insight. The brain is an electrical organ. Its computations are electrical. Its communications are electrical. Its disorders, as we will see in Part Three, are disorders of electrical function. And its potential to be repaired, enhanced, and understood by technology, as we will see in Parts Four and Five, depends entirely on our ability to read, write, and interpret the electrical signals that constitute its activity.

But there is one more thing that the history of Galvani and Volta, of du Bois-Reymond and Helmholtz, of Hodgkin and Huxley, teaches us that is at least as important as any specific scientific finding. It teaches us that the greatest discoveries in science are almost never made by people who set out to make great discoveries. Galvani was trying to understand muscle physiology when he noticed the kicked frog. Helmholtz was trying to test a philosophical hypothesis about the relationship between the mind and the physical world when he measured the speed of the nerve impulse. Hodgkin and Huxley were trying to understand what an action potential actually was before they understood what they needed to do to find out. The history of bioelectricity is a history of curiosity following itself into unexpected places and finding, in those unexpected places, the foundations of an entirely new understanding of what life is. It is not a history of plans. It is a history of attention, of the patient, rigorous, honest attention that science at its best brings to the world, and of the world's repeated willingness to surprise everyone who pays it.

CHAPTER TWO

86 Billion Sparks

The Architecture of the Electrical Brain

LET US BEGIN WITH A number. Not a number in the ordinary sense, the kind that has a definite location on a number line and a clear relationship to the quantities of everyday life. This is a number of a different kind: one of those numbers so far beyond the range of human intuition that it functions more as an image, a gesture toward vastness, than as a quantity in any practical sense. The number is 86 billion, and it represents, as accurately as modern neuroscience can estimate, the count of neurons in a single adult human brain.

Eighty-six billion. You could count one neuron per second, without sleeping, without stopping, without ever making an error, and you would still be counting when the sun expands to swallow the Earth five billion years from now. You could line up all the neurons in a single human brain end to end and the resulting chain would stretch from the Earth to the Moon and back more than 20 times. The neurons in your brain collectively contain enough DNA to fill thousands of encyclopaedias. And each of those 86 billion neurons is connected to thousands of others, creating a network whose total number of connections, somewhere in the range of 100 trillion, exceeds the number of stars in a thousand Milky Way galaxies.

These numbers are genuinely difficult to hold in the mind, which is appropriate, because the mind that is trying to hold them is the very thing they are describing. The brain is trying to comprehend the brain.

The electrical network is trying to model itself. There is something vertiginous about this that I do not want to paper over with the calming language of science writing, because the vertigo is appropriate and it is informative. The human brain is the most complex object that the universe has yet produced, so far as we know, and a proper appreciation of its complexity is not merely a matter of intellectual hygiene. It is the beginning of understanding why consciousness exists at all.

What a Neuron Is and What It Does

Before we can understand the brain's electrical symphony, we need to understand its instruments: the individual neurons that generate and transmit the electrical signals from which everything else is built. A neuron is a cell, but it is a cell of such extraordinary specialization that it differs more radically from a typical body cell than a concert violinist differs from a carpenter. Both are humans, both use similar basic biological machinery, but the specific form that machinery has taken in service of a specific function has transformed it into something barely recognizable as the same thing.

A neuron has, like all cells, a cell body where its nucleus lives and its metabolism is conducted. But it also has two kinds of specialized extensions that no other cell type possesses. The dendrites are branching, tree-like projections that receive incoming signals from other neurons, sometimes thousands of them simultaneously, and funnel those signals toward the cell body. The axon is a single long projection, sometimes several feet long in large animals, that carries the neuron's output signal away from the cell body to its targets, which may be other neurons, muscle cells, or gland cells. The axon is typically wrapped in a fatty insulating sheath called myelin, which dramatically increases the speed at which electrical signals can travel along it, in much the same way that insulation on an electrical wire prevents current from leaking out and allows it to travel faster and further.

The electrical life of a neuron is organized around a single, all-or-nothing event: the action potential. When a neuron receives sufficient input from its dendrites to bring its membrane potential past a threshold value, approximately minus 55 millivolts in most neurons, a cascade of ion channel openings sweeps along the axon at speeds of up to 120 metres per second. This is the action potential. It is, at one level, a simple binary event: either it happens or it does not, either the threshold is crossed or it is not. But at another level, it is the fundamental unit of all neural computation, the dot and dash of the brain's Morse code, the quantum of information in the most complex information-processing system in the known universe.

The all-or-nothing character of the action potential means that neurons cannot vary the intensity of a single signal. They can only vary the rate at which they fire. A neuron responding to a bright light fires more action potentials per second than one responding to a dim light. A motor neuron commanding a forceful muscle contraction fires at higher frequency than one commanding a gentle one. The frequency of firing is the neuron's language, and reading that language, understanding what different patterns of firing mean in different neural circuits, is one of the central challenges of modern neuroscience.

The Synapse: Where Electricity Becomes Chemistry Becomes Electricity Again

When an action potential reaches the end of an axon, it faces a gap. Between the axon terminal of one neuron and the dendrite of the next, there is a space approximately 20 nanometres wide called the synaptic cleft. Twenty nanometres is almost incomprehensibly small: if you enlarged it to the width of a human hair, the neuron would be the size of a city. Yet this tiny gap is one of the most important biological spaces in the known universe, because it is where the electrical signal is converted into a chemical signal and then, on the other side, converted back into electricity.

When an action potential arrives at the axon terminal, it triggers the opening of calcium channels and the influx of calcium ions into the terminal. The calcium triggers the fusion of small membrane sacs called vesicles with the outer membrane of the axon terminal, releasing into the synaptic cleft a collection of chemical molecules called neurotransmitters. These molecules diffuse across the gap in a fraction of a millisecond and bind to receptor proteins on the surface of the receiving neuron's dendrite. Depending on the type of neurotransmitter and the type of receptor, this binding either opens ion channels that drive the receiving neuron's membrane potential toward the firing threshold, making it more likely to generate its own action potential, or it opens channels that push the potential away from threshold, making firing less likely. The first type of synapse is called excitatory, the second inhibitory.

Every moment of your conscious life, millions of these synaptic events are occurring throughout your brain. Right now, as you read this sentence, the neurons of your visual cortex are being activated by the signals arriving from your retina, processed through layers of increasingly abstract representation until the shapes of letters become words become meaning. The neurons of your language areas are interpreting those words in the context of everything you have previously read and known and thought. The neurons of your prefrontal cortex are maintaining your reading goal against the inevitable distractions of a mind that never sits entirely still. The neurons of your emotional centers are coloring the meaning of what you read with whatever affective tone they deem appropriate. And the neurons of your memory systems are connecting what you are reading now to everything relevant that you have ever read before.

All of this is happening electrically, chemically, synaptic event by synaptic event, at a rate and a complexity that makes the most powerful computer ever built look like an abacus. And it is all happening in you, right now, in the 1.4 kilograms of electrochemical tissue that is the most astonishing object in the known universe.

Brain Waves: The Music of the Electrical Mind

If you place electrodes on the surface of a person's scalp and measure the electrical activity of the brain beneath, what you detect is not the firing of individual neurons, which are too small and too numerous for surface recordings to resolve. What you detect is the combined, synchronized electrical activity of millions of neurons firing together in rhythmic patterns: the brain waves that the electroencephalograph records and that neuroscientists have been interpreting, with increasing sophistication, for the better part of a century.

Brain waves are typically described by their frequency, measured in cycles per second or hertz, and they are organized into named bands that correspond to distinct brain states. Delta waves, oscillating at one to four hertz, are the slowest and most powerful and are associated with the deepest stages of non-REM sleep. Theta waves, at four to eight hertz, appear during drowsiness, meditation, and the early stages of sleep, and are also prominent in the hippocampus during memory encoding and spatial navigation. Alpha waves, at eight to thirteen hertz, are the dominant rhythm of a relaxed, wakeful mind with eyes closed, often described as the brain's idling state. Beta waves, at thirteen to thirty hertz, are associated with active, engaged cognition, with focused attention and problem-solving and the processing of new information. And gamma waves, the fastest of the named bands at thirty hertz and above, are associated with intense cognitive processing, perceptual binding, and, in the view of some researchers, the neural correlates of conscious awareness itself.

These rhythms are not merely descriptive labels. They reflect real functional states of the brain, real patterns of communication between brain regions, real processes of computation and integration that matter enormously for everything from how well you learn a new skill to how effectively your brain clears its metabolic waste products during sleep. When the brain's rhythms are disrupted, as they are in epilepsy, in depression, in schizophrenia, in Alzheimer's disease, the consequences for neural function are severe and often devastating. When the rhythms are properly coordinated, as they are in deep sleep, in meditative states, in the flow experiences of creative and athletic performance, the brain operates at its most efficient and most effective.

Perhaps the most important of these rhythms, and the one that has attracted the most research attention in the past two decades, is the gamma oscillation. Gamma waves are generated when large populations of neurons synchronize their firing at high frequency, creating a rapid, coherent rhythm that some neuroscientists believe is essential for binding together the various aspects of a perceptual experience into a unified whole. When you see a red ball rolling across a green lawn, the redness and the roundness and the motion and the greenness and the spatial relationships between these features are all processed by different populations of neurons in different regions of your visual cortex. The gamma oscillation, in this view, is the mechanism by which these separately processed features are bound together into the single, unified experience of a red ball rolling across a green lawn. This is sometimes called the binding problem, and it is one of the deepest unsolved problems in all of neuroscience, sitting right at the edge of the Hard Problem of consciousness itself.

The Brain Is Not One Brain

One of the most persistent and most misleading images of the brain is that of a single, unified organ, a gray walnut-shaped mass that does brain things in the way that a heart does heart things or a kidney does kidney things. The reality is far more interesting and far more strange. The human brain is not one brain. It is a collection of anatomically and functionally distinct regions, each with its own architecture, its own electrical character, its own connectivity profile, and its own specialization for particular computational tasks, all integrated into a dynamic, constantly changing whole through an intricate web of long-range connections.

The cerebral cortex, the deeply folded outer layer of the brain that is most visible in any image of the human brain, contains about 16 billion of the brain's 86 billion neurons and is organized into distinct lobes: the occipital lobe at the back, primarily devoted to visual processing; the temporal lobes on the sides, crucial for language, memory, and the recognition of objects and faces; the parietal lobes above and behind, involved in spatial processing, attention, and the integration of information from different sensory modalities; and the frontal lobes at the front, which are the most recently evolved regions of the human brain and are responsible for the highest-order cognitive functions: planning, reasoning, impulse control, working memory, and the integration of information from all other brain regions into goal-directed behavior.

Below the cortex, the subcortical structures perform functions that are in many ways more ancient and more fundamental. The hippocampus, a curved structure in the medial temporal lobe, is essential for the formation of new declarative memories, as the tragic case of Henry Molaison would later demonstrate. The amygdala, almond-shaped and located just in front of the hippocampus, is the brain's alarm center, generating fear responses and tagging experiences with emotional significance. The basal ganglia, a collection of nuclei deep in the brain, are crucial for the learning and execution of motor sequences and for the reward-based learning that shapes habitual behavior. The cerebellum, at the back of the brain beneath the cortex, contains more neurons than all the rest of the brain combined and is essential for the coordination and fine-tuning of movement. And the brainstem, the most evolutionarily ancient part of the brain, controls the basic life-sustaining functions: breathing, heart rate, blood pressure, arousal, and the fundamental regulation of sleep and wakefulness.

What makes the human brain distinctively human, to the extent that anything does, is not the presence of any of these structures in isolation, all of which are shared with other mammals. It is the extraordinary development of the prefrontal cortex, the extraordinary density and complexity of the connections between brain regions, and perhaps above all the extraordinary capacity for flexible, context-sensitive integration of information across time and space that these connections support. The human brain can hold a plan in mind across hours or days, working toward a distant goal against the immediate pull of competing impulses. It can use language to represent abstract categories and logical relationships that have no direct perceptual referent. It can simulate alternative futures, evaluate their probable consequences, and choose between them on the basis of values that it has itself constructed and can itself revise. These capacities are the basis of human civilization, of science, of art, of moral reasoning, of everything that distinguishes the human historical record from the behavioral repertoire of any other animal. And they all run on electricity.

The Power of the Sleeping Machine

There is one more aspect of the brain's electrical architecture that deserves mention here, both because it is often overlooked in popular accounts and because it will become central to the chapter on sleep later in this book. The brain is extraordinarily expensive, metabolically speaking. Although it accounts for only about two percent of the body's mass, it consumes approximately twenty percent of the body's total energy budget at rest. This energy is used almost entirely to maintain the ionic gradients that make the action potential possible: the sodium-potassium pump, which constantly expels sodium ions from the cell interior and pumps potassium ions in, runs continuously and consumes more ATP than almost any other process in the body.

This metabolic cost has consequences. The brain's intensive metabolism produces waste products that must be cleared. For most of the body's tissues, waste products are cleared through the lymphatic system, the network of vessels that drain fluid from tissues back into the circulation. For decades, neuroscientists were puzzled by the fact that the brain appeared to have no lymphatic system and no obvious mechanism for waste clearance. The puzzle was solved in 2013, when the neuroscientist Maiken Nedergaard and her colleagues at the University of Rochester discovered the glymphatic system: a brain-specific waste clearance mechanism that uses the channels surrounding blood vessels as a drainage system, pumping cerebrospinal fluid through the brain tissue to flush out metabolic waste products.

The crucial insight, which we will explore in depth in the chapter on sleep, is that the glymphatic system operates primarily during sleep, and specifically during the slow-wave sleep that is associated with the brain's delta rhythms. During wakefulness, the channels of the glymphatic system are relatively constricted. During sleep, they expand, increasing the flow of cerebrospinal fluid through the brain by as much as sixty percent and dramatically accelerating the clearance of waste products including, most significantly, amyloid-beta protein, the molecule that accumulates in the brains of Alzheimer's patients and is now widely believed to play a central role in the disease's progression.

This is why sleep deprivation is so damaging, and why the science of sleep has become one of the most active areas of brain research in the past decade. The brain's electrical rhythms during sleep are not a sign of reduced activity. They are the signature of a brain performing operations of maintenance and consolidation that are just as essential as anything it does while awake, and that, if neglected, lead to exactly the kind of electrochemical dysfunction that is the subject of Part Three of this book.

The Question That Will Not Go Away

We have now covered the basic architecture of the brain's electricity: the neurons, the action potentials, the synapses, the brain waves, the regional organization, the metabolic demands. This is the foundation on which everything else in this book is built. But before we leave this chapter, I want to return to the question that was raised in the preface and that will haunt every subsequent chapter: the question of how all of this electricity becomes experience.

Consider what we have established. The redness of red is processed by specific populations of neurons in the V4 region of your visual cortex, which respond selectively to wavelengths of light in the range of 620 to 750 nanometres. The sound of a middle C played on a piano is processed by neurons in the primary auditory cortex that respond to sound waves oscillating at 261 hertz. The pain of a burned finger is processed by nociceptive neurons in the somatosensory cortex that receive signals from pain receptors in the skin. The memory of your grandmother's kitchen is stored in a distributed network spanning the hippocampus, the temporal cortex, the olfactory cortex, and several other regions, encoded in the specific pattern of synaptic strengths connecting those neurons.

All of this is electrical. All of it is mechanistically described. And yet: when you see red, you do not merely process 700-nanometre wavelengths. You experience redness, a quality of awareness that has a specific character, a specific feel, that is not contained in any description of the wavelength or the neural response to it. When you hear middle C, you experience a sound that has a particular tone, a particular warmth or brightness, a particular relationship to the other sounds around it. When you feel pain, you experience something that is not just a signal but a suffering. When you remember your grandmother's kitchen, you experience something that feels like presence, like reaching back through time and touching something real.

Where does that quality come from? Where does the experience go? None of the mechanisms described in this chapter, as accurate and as important as they are, contain an answer. The neurons fire. The ions flow. The synapses open and close. The brain waves rise and fall. And somehow, out of all of this, there is something it is like to be you. That is the mystery. And it is to understanding that mystery, and to appreciating why it is both the hardest and the most important problem in all of science, that the next chapter turns.

CHAPTER THREE

The First Explorers of the Electric Mind

From Electroencephalography to the Neuroscience Revolution

HANS BERGER WAS AFRAID of death. Not in the ordinary way that most people are afraid of death, as an abstract future event to be contemplated occasionally and then set aside. He was afraid of it in the specific, urgent, immediate way of a man who believed he had once come very close to it through a connection he could not explain, a connection that he believed was telepathic in origin, and that he spent the rest of his life trying to understand through science.

The incident happened in 1892, when Berger was a young cavalry soldier on manoeuvres near the German town of Würzburg. His horse stumbled on a steep hillside and he was thrown forward into the path of an artillery piece, where he was almost killed. He survived by a margin that seemed, to him, miraculous. That same evening, he received a telegram from his father in response to a letter that his father had not yet received: a letter in which Berger's sister had written to their father describing a powerful premonition that Hans was in danger. The timing, the sister's premonition and the artillery horse and the near-miss all coinciding within the same few hours, struck the young Berger as too significant to be coincidence. He became convinced that he had transmitted his fear and danger to his sister across the distance, through some kind of mental force that physical science had not yet identified.

He became a psychiatrist. He became a scientist. And he devoted the next four decades of his life to understanding the brain's electrical activity, not primarily because he was interested in the clinical applications of brain measurement, though those interested him too, but because he believed that the brain's electricity might be the physical substrate of the mental force he had sensed in that cavalry field in 1892. He was looking for the physics of the mind. What he found, instead, was something far more practically important and scientifically enduring.

A Hundred Years of Skepticism

The story of Hans Berger's invention of the electroencephalograph is one of the most remarkable stories in all of twentieth-century science, not only for the importance of what he discovered but for the extraordinary isolation and skepticism under which he discovered it. For more than a decade, Berger worked alone, in almost total secrecy, not publishing a single word about his research, because he was terrified of being dismissed and because he wanted to be absolutely certain before he made a claim that he knew would be regarded as extraordinary.

His experimental setup was simple in principle and brutally difficult in practice. He placed electrodes on the scalp of a human subject, connected them to a galvanometer of the kind that du Bois-Reymond had developed in the previous century, and attempted to record the tiny electrical fluctuations produced by the neural activity of the brain beneath. The signals were almost impossibly small: thousandths of a volt, embedded in a sea of noise from muscle activity, electrical interference from building wiring, and the limitations of his recording equipment. For years, Berger was unable to produce recordings clear enough to be interpretable.

His breakthrough came on July 6, 1924, when he placed electrodes on the scalp of his fifteen-year-old son Klaus, who had volunteered to be a research subject, and recorded what appeared to be a clear, rhythmic electrical oscillation of approximately ten cycles per second, present when Klaus was resting with his eyes closed and disappearing when he opened them and began to look at something. Berger recognized immediately, even if he could not yet be certain, that he was seeing the electrical activity of the brain itself: what we now call the alpha rhythm, the signature of relaxed, wakeful awareness.

He kept working, refining his techniques, extending his recordings to more subjects, and eventually, in 1929, he published the first of fourteen papers on what he called the electroencephalogram, the EEG. The title of the first paper, On the Human Electroencephalogram, was modest almost to the point of invisibility, and its reception was correspondingly muted. The German scientific establishment largely ignored it. The British establishment, whose opinion carried enormous weight in international science at the time, was more openly dismissive. Berger was not, after all, a major figure in mainstream physiology. He was a psychiatrist at a relatively provincial German university, working on equipment that most physiologists would have considered inadequate for the task he claimed to have accomplished. And the claim he was making, that one could record the electrical activity of the brain from the outside of the skull, seemed to many experts simply implausible on technical grounds.

Vindication from an Unlikely Quarter

The vindication came in 1934, from two British scientists who were among the most respected physiologists of their generation: Edgar Douglas Adrian and Bryan Harold Cabot Matthews. Adrian was already a Nobel laureate, having won the prize in 1932 for his work on the function of nerve fibers, and his endorsement of Berger's findings carried the kind of authority that no amount of careful German skepticism could dismiss. Adrian and Matthews obtained Berger's published records, replicated his experiments on their own equipment, and confirmed that the electrical oscillations Berger had recorded were real, reproducible, and genuinely of neural origin. They announced their confirmation at a meeting of the Physiological Society in London in 1934, and the scientific establishment listened.

The EEG moved rapidly into clinical use in the years that followed. British and American neurologists recognized almost immediately that it could be used to diagnose epilepsy, a condition in which the brain's electrical abnormalities produced distinctive and readily identifiable patterns in the EEG record. The diagnosis of epilepsy had previously depended entirely on the clinical history and eyewitness accounts of seizure episodes, which were often ambiguous and sometimes completely absent if the patient had no witnesses. The EEG provided, for the first time, an objective biological marker of the condition, and it transformed the management of epilepsy in ways that are still felt in neurology clinics today.

Berger himself did not live to see the full flowering of what he had started. He was an unwilling collaborator with the Nazi regime that came to power in Germany the year before his vindication, and he was eventually forced into retirement by the Nazi authorities in 1938, partly because of his refusal to engage with the regime's ideology and partly because of the interference of a professional rival. He died by suicide in 1941, a deeply depressed man who felt that the world had failed to understand the most important work of his life. He was right that the world had been slow to understand. He was wrong that it had failed. The EEG is still the most widely used neurological diagnostic tool in the world a century after his first recording of it, and it has been joined by a family of descendant technologies that have collectively transformed the study of the human brain beyond anything Berger could have imagined in his lonely laboratory in Jena.

Wilder Penfield and the Electric Map of the Mind

While Berger was working in isolation in Germany, a Canadian neurosurgeon named Wilder Penfield was conducting a series of experiments in Montreal that would prove to be among the most revealing and, to many of his subjects, among the most personally extraordinary investigations ever performed on the living human brain. Penfield was one of the leading figures in the surgery of epilepsy, developing techniques for removing the brain tissue responsible for generating seizures while preserving as much healthy tissue as possible. To do this safely, he needed to know exactly where in a given patient's brain the critical functions of language, movement, sensation, and memory were located, because these locations vary somewhat from person to person and cannot be reliably predicted from anatomical landmarks alone.

His solution was elegant, effective, and deeply unusual: he operated on his patients under local anaesthesia, with the patients awake and able to communicate throughout the procedure. With the skull open and the brain surface exposed, he used a small electrical probe to stimulate individual points on the cortex and asked the patient to report what they experienced. A stimulus to one point might cause the patient's hand to move involuntarily. A stimulus to another might produce a tingling sensation in the face. A stimulus to the speech areas might cause the patient to be unable to find words, even though they could clearly hear and understand what was being said to them.

These mapping procedures produced the somatotopic maps of the human brain that are still reproduced in every neuroscience textbook: the motor homunculus and the sensory homunculus, distorted cartoon figures in which the size of each body part reflects the amount of cortical territory devoted to its control or sensation. The hand, the face, and the tongue are enormously enlarged relative to their actual physical size because they are represented by disproportionately large areas of cortex, reflecting the extreme precision of the motor control and sensory discrimination that these regions require.

But Penfield's most startling discoveries came when he stimulated the temporal lobes, the regions of cortex on the sides of the brain that are involved in memory, language, and the processing of complex sensory experiences. When he stimulated certain points in the temporal lobes of epileptic patients during surgery, the patients reported experiences unlike anything produced by stimulation of the primary sensory or motor cortices. They did not feel a tingling sensation or see a flash of light or experience an involuntary movement. They heard music. They heard voices. They smelled smells. And, most remarkably of all, they experienced vivid, detailed memories that felt as real and as present as the operating room around them.

One patient heard an orchestra playing a piece of music she recognized from her childhood, so clearly that she could hear individual instruments and describe the emotional character of the performance. Another experienced herself as a young girl in her backyard, watching her brother playing with a dog, hearing the sounds of the garden around her with the full sensory richness of a real memory. Another heard the voice of his dead mother calling his name from another room. These experiences were not like remembering in the ordinary sense: they were not the pale, reconstructed, often fragmentary quality of normal memory. They were, as Penfield's patients consistently described, indistinguishable from being there.

What Penfield had discovered, by accident, was that memory is not stored in a generalized, diffuse way throughout the brain but is at least partly associated with specific anatomical locations in the temporal cortex. A tiny electrical current, delivered to the right point, could reproduce a specific memory with a vividness that normal voluntary recall could not match. The implication, which Penfield himself was cautious about drawing too strongly, was that memories are not so much retrieved as re-experienced: that remembering is not reading from a filing system but re-playing an electrical pattern that was originally laid down by the experience being remembered.

The Sleep Discovery That Changed Everything

In the autumn of 1952, a graduate student at the University of Chicago named Eugene Aserinsky was conducting a study of infant sleep, monitoring the eye movements of sleeping babies as part of an investigation into the relationship between eye activity and brain states. He noticed something that had been seen before by other researchers but never adequately explained: during certain periods of sleep, the eyes of his subjects moved rapidly back and forth under their closed lids, as if watching something happening in front of them. This rapid eye movement, which occurred at predictable intervals throughout the night, was associated with a distinctive EEG pattern: one that looked almost identical to the EEG of a waking, alert brain.

Aserinsky brought his observation to his supervisor, Nathaniel Kleitman, who was the world's leading authority on sleep research and who initially received the report with some skepticism. The idea that there were distinct stages of sleep with radically different electrical characteristics was not new, but the specific observation that one sleep stage produced brain waves essentially indistinguishable from wakefulness was surprising enough to deserve careful investigation. Aserinsky and Kleitman studied the phenomenon systematically, monitoring adult subjects through the night and waking them during episodes of rapid eye movement to ask what they had been experiencing. The subjects consistently reported vivid, narratively coherent dreams, far more elaborate and more visually detailed than the fragmentary, confused experiences reported by people woken from other sleep stages. The association between rapid eye movements and dreaming was established, and the sleep stage responsible was given the name that it still carries: rapid eye movement sleep, or REM.

The discovery of REM sleep transformed the scientific understanding of sleep from a state of passive rest to an active, electrically complex process involving at least two fundamentally different kinds of brain activity. It also opened the scientific study of dreaming to rigorous empirical investigation for the first time, because the REM signature gave researchers a reliable way to identify when dreaming was occurring and to study its physiological correlates in systematic ways. In the decades that followed, sleep researchers would map the full architecture of the sleep cycle, the characteristic alternation between non-REM stages of increasing depth and periodic episodes of REM, that repeats approximately four to six times during a full night's sleep, and would begin to understand the distinct functional roles of each stage in memory consolidation, emotional regulation, and brain maintenance.

The Modern Imaging Revolution

The EEG that Berger invented and the maps that Penfield drew gave neuroscience its first ways of seeing the living human brain in action. But they were limited in important ways. The EEG records electrical activity with extraordinary time resolution, tracking the brain's dynamics millisecond by millisecond, but its spatial resolution is poor: the electrical signals detected by scalp electrodes reflect the combined activity of millions of neurons spread over large areas of cortex, and it is difficult to determine precisely where in the brain a given signal is coming from. Penfield's electrical stimulation maps, on the other hand, were spatially precise to the millimetre but required open-skull surgery and could only be performed in the course of medically necessary procedures.

The resolution of this dilemma came gradually, through a series of technological innovations that transformed neuroscience in the last quarter of the twentieth century. The most important of these was functional magnetic resonance imaging, or fMRI, which was developed in the early 1990s and which measures changes in blood oxygenation as a proxy for neural activity. When a brain region becomes more active, its oxygen consumption increases and blood flow to that region increases to match, changing the magnetic properties of the blood in ways that the MRI scanner can detect. The result is a three-dimensional map of brain activity that can be resolved to approximately a cubic millimetre, far better spatial resolution than EEG, acquired while the subject lies comfortably in a scanner without any surgical intervention.

The limitation of fMRI, conversely, is its temporal resolution: because it tracks blood flow changes rather than electrical activity directly, it responds relatively slowly to neural events, with a time resolution of several seconds rather than the milliseconds of EEG. The two technologies are thus complementary: EEG tells you when the brain is doing something; fMRI tells you where. Combining them, as researchers increasingly do, gives a more complete picture than either technique alone can provide.

More recently, the development of high-density EEG systems with 128, 256, or more electrodes, combined with sophisticated mathematical algorithms for source localization, has allowed researchers to significantly improve the spatial resolution of scalp recordings, approaching in some conditions the spatial precision of fMRI while retaining the temporal precision that is EEG's fundamental advantage. And entirely new techniques continue to emerge: magnetoencephalography, which measures the tiny magnetic fields generated by neural currents and combines high spatial and high temporal resolution in a single measurement; single-photon emission computed tomography, which tracks blood flow using radioactive tracers; and the various forms of optical brain imaging that use near-infrared light to measure changes in the optical properties of brain tissue associated with neural activity.

Together, these technologies have produced an explosion in our knowledge of the brain's functional organization that has no parallel in the history of neuroscience. In the past three decades, we have mapped the neural correlates of language and attention, of emotion and memory, of social cognition and moral reasoning, of creativity and decision-making, with a level of detail and reliability that would have seemed impossible to Berger working alone in his Jena laboratory with his galvanometer and his son Klaus. We have identified the brain regions that are specifically disrupted in depression, in schizophrenia, in autism, in Alzheimer's disease. We have traced the development of the brain from infancy through adolescence into adulthood, mapping the changes in connectivity and functional organization that accompany the emergence of mature cognitive and emotional capacities. We have begun to understand how the brains of people who meditate, who play musical instruments, who speak multiple languages, who have undergone specific kinds of trauma, differ from those of people who have not had these experiences, and to identify the neural mechanisms by which experience shapes the brain's architecture throughout the lifespan.

This is the neuroscience revolution that the first explorers of the electric mind made possible: Berger with his lonely EEG recordings, Penfield with his surgical maps, Aserinsky and Kleitman with their sleeping subjects and their overnight monitors. They were working in the dark, with instruments that now seem almost touchingly primitive, guided by intuitions that were often partially wrong and occasionally entirely right. They gave us the first real glimpse into the electrical life of the human brain. Everything that has come since is built on the foundations they laid.

PART TWO

THE ELECTRICAL ARCHITECTURE OF THE MIND

How Electricity Creates Memory, Emotion, Sleep, and Consciousness

Part Introduction

The Most Personal Electricity

Part One of this book was concerned with the physical and historical foundations: what bioelectricity is, how it was discovered, how the action potential was mathematically described, how the first tools for observing the living brain's electrical activity were developed. The chapters of Part One were, at their core, chapters about mechanisms: the machinery of the electric brain, explained as clearly and as completely as the science allows.

Part Two is different in character. It is still concerned with mechanisms, because mechanisms are what science has to offer and they are genuinely extraordinary. But in Part Two, the mechanisms are in service of questions that are more personal, more immediate, and for many readers more urgently relevant than the historical and technical questions of Part One. How does the brain's electrical activity produce memory? How does it create the emotions that color every experience of conscious life? What is actually happening in the brain during sleep, and why does what happens there matter so much for everything else? And above all: how does any of it produce consciousness, the subjective, first-person experience of being a self, of there being something it is like to be you?

These are the questions of Part Two. They are not questions to which science has complete answers. In the case of consciousness in particular, science has not yet even agreed on how to approach an answer, let alone arrived at one. But the partial answers that have been developed over the past century of neuroscience are remarkable enough to justify every page that follows, and the questions themselves, held clearly in the mind and taken seriously in their full difficulty, are among the most interesting that any human being has ever asked.

A word about the writing style of Part Two. The chapter on memory, Chapter Four, is written in the second person throughout: it addresses you, the reader, directly, walking you through your own memory processes as they are described. This is a deliberate choice. Memory is not an abstract system. It is the architecture of your personal identity, the mechanism by which you remain the same person across time. Writing about it in the third person would create a distance that does not exist between you and your memories. The intimacy of the second person is an attempt to honor that.

CHAPTER FOUR

The Memory Machine

How Electricity Writes, Stores, and Rewrites the Story of Your Life

YOU ARE DOING IT RIGHT now. Not reading, though you are also doing that. You are making memories. Every word your eyes pass over, every sentence that assembles itself into meaning in your mind, every idea that connects to something you have previously known or thought or felt: all of this is being encoded, to varying degrees of permanence, in the electrical and chemical structure of your brain. Most of what you are reading will be gone by tomorrow. Some fragments will survive a week. A very small number of ideas, the ones that genuinely surprise you, that contradict something you believed, that connect to something you care about deeply, will be with you for years, perhaps for the rest of your life. You cannot fully predict which will be which. Neither can I. But the process by which your brain is deciding, right now, what to keep and what to discard, is one of the most fascinating stories in all of neuroscience, and it is a story about electricity.

The Case That Changed Everything

His name was Henry Gustav Molaison. For most of his life, he was known to the scientific world only as H.M., a precaution taken to protect his privacy that was maintained until his death in 2008. He was born in Connecticut in 1926 and had a normal childhood until, at the age of nine, he suffered a head injury that may have contributed to the development of severe epilepsy. By his mid-twenties, his seizures had become so frequent and so debilitating that he could barely function: he was having up to eleven major convulsive episodes per day, and none of the medications available in the early 1950s provided adequate control.

In 1953, Henry came under the care of the neurosurgeon William Beecher Scoville, who proposed a radical treatment: the surgical removal of the medial temporal lobes, including the hippocampus on both sides of Henry's brain, which were believed to be the source of his seizures. The surgery was performed at Hartford Hospital on September 1, 1953. It worked remarkably well for the epilepsy. Henry's seizure frequency dropped dramatically, and the severity of those that remained was greatly reduced. But as the days and then weeks after surgery passed, it became clear that the operation had produced a consequence that no one had anticipated and that would make Henry Molaison's case one of the most important in the history of neuroscience.

Henry could no longer form new long-term memories. Not simply reduced in memory capacity, or impaired in some modalities and not others, but fundamentally, categorically unable to retain new declarative memories, the kind of memory that allows you to know things and to know that you know them, for more than a few minutes. He remembered everything that had happened before his surgery, his childhood, his family, his school, his town, with reasonable accuracy, though his recall of the years immediately before the surgery was somewhat patchy, a phenomenon that would later be recognized as retrograde amnesia and that indicated that the hippocampus is not just the site of new memory formation but also plays a role in the consolidation of recently formed memories. But after his surgery, every new experience, every new face, every new fact, every conversation he had with the neuropsychologist who would study him for the next fifty-five years, evaporated within minutes, leaving no trace.

He woke every morning not knowing what year it was. He consistently estimated his own age as close to twenty-seven, the age he had been when his surgical memory ended. He recognized photographs of people who had been famous before 1953 and did not recognize photographs of people who had become famous after it, even people who had become extremely well known internationally. Every morning, the neuropsychologist Brenda Milner, who worked with Henry for decades at the Montreal Neurological Institute, was a stranger to him, even though he might have spent several hours with her the previous day. When he was told that his father had died, he was shocked and distressed. The next day, he would be told again, and would be shocked and distressed again, because the memory of being told had not been retained. He was, in a sense, permanently marooned in 1953, at the age of twenty-seven, while the world and everyone around him aged and changed.

What Henry's tragic case proved, more clearly than any planned experiment could have designed, was that the hippocampus is not the seat of all memory, as some had speculated, but is specifically and critically necessary for the formation of new declarative memories: the memories of facts and events that can be consciously recalled and verbally reported. It also proved that other forms of memory are mediated by different brain systems, because Henry retained the ability to learn new motor skills. Taught to trace the outline of a star while looking at his hand in a mirror, a difficult task that requires coordinating movements with reversed visual feedback, he improved his performance over repeated sessions just as a normal person would. And when shown the tracing task each time, he reported having no memory of ever doing it before. He was learning. He just did not know that he was learning. The skill was being encoded in his motor systems, in the cerebellum and the basal ganglia, brain structures that his surgery had left intact.

The Electrical Machinery of Memory Encoding

Your hippocampus is working constantly, though you almost never have reason to think about it. As you move through your day, encountering new information, new experiences, new faces and places and ideas, your hippocampus is receiving input from virtually every sensory and cognitive system in your brain and is doing something with that input that no other brain structure can do: it is binding together the various elements of an experience into a unified, relational memory that can be retrieved later as a whole.

The mechanism by which this binding and encoding occurs was discovered, in its essential form, by the British neuroscientist Timothy Bliss and his Norwegian colleague Terje Lømo, working at the University of Oslo in the early 1970s. They were studying the hippocampus of anaesthetized rabbits, recording the electrical responses of hippocampal neurons to stimulation of their inputs, when they discovered something that neither had predicted: if you stimulate a hippocampal input pathway at high frequency for a brief period, the subsequent responses of the neurons in that pathway to single stimuli are dramatically increased, and this increase persists for hours, days, or even weeks after the original high-frequency stimulation. The pathway had been strengthened by its activity. It had, in the language that Bliss and Lømo gave to this phenomenon, undergone long-term potentiation.

Long-term potentiation, or LTP, is now understood to be the principal cellular mechanism of learning and memory formation in the brain. When two neurons are active simultaneously, or when a presynaptic neuron fires repeatedly in rapid succession, the synapse between them undergoes molecular changes that increase its strength: more receptor proteins are inserted into the postsynaptic membrane, more neurotransmitter is released from the presynaptic terminal, or structural changes occur that increase the physical size of the synapse. The result is that the same presynaptic signal produces a larger postsynaptic response than it did before the potentiation. The synapse has been changed by its activity, and that change is the physical substrate of the memory.

The molecules involved in this process have been worked out in extraordinary detail over the past fifty years. The key receptor in LTP is the NMDA receptor, a type of glutamate receptor that acts as a molecular coincidence detector: it only opens its ion channel when the presynaptic neuron is releasing glutamate and the postsynaptic neuron is already depolarized, meaning that both neurons must be active at the same time for the receptor to be activated. This is the molecular implementation of what is sometimes called Hebb's rule, after the Canadian psychologist Donald Hebb, who proposed in 1949 that neurons that fire together wire together: that simultaneous activity in connected neurons strengthens the synaptic connections between them. When the NMDA receptor opens, it allows calcium to flow into the postsynaptic cell, and it is this calcium influx that triggers the cascade of molecular events, the activation of kinases, the insertion of AMPA receptors, the growth of new dendritic spines, that constitute the physical change that underlies the memory.

Why Memory Is Not a Recording

Here is something that the science of memory makes very clear, something that is both fascinating and, once you truly understand it, somewhat unsettling: your memories are not recordings. They are not files stored on a biological hard drive that can be retrieved intact and played back. They are reconstructions, rebuilt from component parts every time you access them, and this reconstruction process is imperfect in ways that have profound implications for how much you can trust your own recollections of the past.

The evidence for this conclusion comes from multiple directions. The psychologist Elizabeth Loftus has spent more than four decades demonstrating, in a series of elegant and sometimes disturbing experiments, that human memory is not merely fallible but actively malleable: it can be altered, distorted, and even completely fabricated by subsequent experiences and suggestions, without the person who holds the memory having any awareness that this has occurred. In one of her most famous paradigms, subjects who witnessed a simulated automobile accident were later asked questions about what they had seen. Subjects who were asked how fast the cars were going when they smashed into each other gave higher speed estimates than subjects asked how fast they were going when they hit each other. And a week later, subjects who had been asked about smashing were more likely to report having seen broken glass in the video, even though there was no broken glass in the video.

The memory was changed by the word used in the question. Not consciously, not deliberately, not even detectably by the person whose memory was changed. The reconstruction process, which is biased toward narrative coherence and toward incorporating information that arrives after the original experience, had quietly revised the record. This is not a bug in the memory system. It is, in an important sense, a feature: a memory system that updated its representations in the light of new information would be more adaptive than one that preserved every detail of every past experience unchanged forever. But it means that the confident sense of accuracy that accompanies vivid memories is no guarantee of actual accuracy. Your memories feel real. They feel accurate. And they are, in important respects, both of those things. They are also, in ways that are often impossible to detect from the inside, reconstructed, interpreted, and sometimes subtly or not so subtly wrong.

The Molecular Reconsolidation Window

The malleability of memory is not just a quirk of the encoding or retrieval process. It is built into the basic biology of how memories are maintained over time. After a long-term memory is initially formed and consolidated, it enters a stable state maintained by the structural changes at synapses that LTP has produced. For many years, neuroscientists assumed that once a memory was consolidated, it was fixed: the molecular changes that maintained it were permanent, and the only way to alter a consolidated memory was through the slow process of new learning.

This assumption was overturned in 2000 by a paper published in Nature by Karim Nader, Glenn Schafe, and Joseph LeDoux at New York University. They trained rats to fear a particular sound by pairing it with mild electric shock, a classical fear conditioning paradigm that depends on the amygdala and produces a strong, durable fear memory. They then reactivated the memory by exposing the rats to the sound alone, and at the moment of reactivation, they infused into the amygdala a drug that blocks protein synthesis, a process essential for the consolidation of new memories. The result was striking: the reactivated memory was erased. The rats that had received the protein synthesis inhibitor at the moment of memory reactivation no longer showed fear to the sound, even though rats that had not had their memory reactivated retained the fear as strongly as ever.

What Nader and his colleagues had demonstrated was that memories do not simply persist in a stable molecular state once they are consolidated. Each time a memory is reactivated, it enters a temporarily labile state in which it must be re-consolidated using the same molecular machinery that was required for its initial formation. This process, called reconsolidation, opens a window of vulnerability during which the memory can be altered or disrupted. The implications of this discovery for understanding memory disorders, for developing treatments for PTSD and phobias, and for the broader question of the reliability of memory, are still being worked out. But it has fundamentally changed the scientific understanding of memory as a process rather than a state: not a fixed record but an ongoing biological activity, perpetually reconstructing itself each time it is accessed.

The Geography of Memory

Memory is not localized in a single brain structure or even in a single type of brain structure. It is distributed across the brain in a way that reflects the different types of information that need to be encoded and the different timescales over which that information needs to be retained. Understanding this geography is essential for understanding both normal memory function and the specific patterns of memory impairment that result from different kinds of brain damage.

The hippocampus, as established by Henry Molaison's case and confirmed by decades of subsequent research, is essential for the initial encoding of declarative memories: the memories of facts and events that can be consciously recalled and verbally expressed. But the hippocampus is not the permanent storage site for these memories. Over a period of months to years after initial encoding, declarative memories undergo a process of consolidation during which their representation gradually shifts from the hippocampus to the neocortex, the outer layer of the brain, where they eventually become independent of the hippocampus and can be retrieved even if the hippocampus is damaged. This is why patients with hippocampal damage, like Henry Molaison, can remember remote memories from the distant past but cannot form new ones: the remote memories have already been consolidated in the neocortex, but the hippocampus required for new memory formation is no longer functioning.

The amygdala, the brain's emotional center, adds an additional layer to memory encoding by enhancing the consolidation of emotionally significant experiences. When an experience is accompanied by strong emotion, the amygdala releases stress hormones that act on the hippocampus to strengthen the encoding of that experience, making it more likely to be retained long-term. This is why emotionally charged memories, both positive and negative, tend to be more vivid and more durable than emotionally neutral ones. It is also why traumatic memories are sometimes encoded with pathological intensity, a phenomenon that underlies the intrusive, vivid, difficult-to-extinguish traumatic memories of PTSD.

The cerebellum and the basal ganglia handle the encoding of procedural memories: the motor skills and habits and reflexes that, like Henry's star-tracing ability, can be learned and retained even when the hippocampal system is destroyed. These procedural systems are also responsible for the kind of conditioning that Pavlov famously studied: the learned associations between neutral stimuli and emotionally significant ones that shape behavior in ways that can persist long after the original conditioning. And the neocortex itself, in its various regions, stores the semantic memories of facts about the world, the episodic memories of specific past events after their consolidation from the hippocampus, the perceptual memories of specific sights and sounds and smells, each in the regions specialized for processing the relevant type of information.

You carry all of this within you, right now: this distributed, reconstructive, imperfect, extraordinarily rich memory system that makes you who you are. Every experience you have had, every person you have known, every idea you have entertained, every skill you have practiced is represented somewhere in the electrical and molecular architecture of your brain. Not perfectly. Not permanently. Not without distortion or decay. But represented, in the patterns of synaptic strength across your neural networks, in the molecular signatures at your synapses, in the connectivity of your cortical regions, that collectively constitute the physical substrate of your autobiography. Your memories are you. And they are electricity.

CHAPTER FIVE

The Emotional Circuit

How Your Brain's Electricity Creates Love, Fear, Grief, and Joy

BEFORE THE SCIENCE, let me give you a moment. Think of the last time you were genuinely afraid. Not mildly apprehensive, not intellectually concerned about something that could go wrong, but truly, physically afraid: heart hammering, breath shortened, muscles tensed, the world narrowed suddenly to the single point of the threat. Now think of the last time you felt genuine joy, the uncomplicated, flooding kind that arrives without warning when something you had hoped for suddenly becomes real, or when a child laughs, or when a piece of music perfectly captures something you had felt but never been able to name. And think of grief. If you have been fortunate enough not to know grief yet, you will. It is the price of love, and it is one of the most devastating and most human experiences available to us.

These experiences, fear and joy and grief and love and rage and tenderness and shame and pride and wonder, are the most personally significant things that happen to us. They are what we mean, at the deepest level, when we talk about a life having been lived fully or partially, happily or unhappily, with meaning or without it. And for most of human history, they were understood as the province of the heart or the soul or the spirit, as experiences that transcended the merely physical, that could not in principle be reduced to the mechanics of biology without losing the very quality that made them what they were.

The neuroscience of the past four decades has not destroyed this understanding. What it has done is something more interesting: it has shown that the emotions are not despite the brain, not in spite of its biological machinery, but are the brain's most sophisticated products. That fear is a computation as real as any other neural computation, implemented in specific circuits, following specific rules, with specific consequences for attention, memory, and behavior. That love involves brain regions and neurotransmitters that we can name and study, without that naming and studying making the experience of love any less real or any less extraordinary. The biology of emotion is not the enemy of the experience of emotion. It is the explanation of how something so complex, so powerful, so distinctively human could exist at all.

The Alarm in the Brain

At the center of the brain's emotional processing is a small, almond-shaped structure buried in the medial temporal lobe, one on each side of the brain, called the amygdala. The name comes from the Greek word for almond, and the structure earns its name not only in shape but in character: compact, dense, surprisingly powerful for its size, and in the right circumstances, capable of producing effects far disproportionate to what its modest dimensions would suggest.

The amygdala is the brain's alarm system. It is the structure that evaluates incoming sensory information for potential threats and generates the fear response when a threat is detected. It does this work with a speed and an automaticity that is the product of millions of years of evolutionary pressure: the ability to detect and respond to danger before it has had time to be consciously analyzed is, in an environment containing predators and competitors, a survival advantage of the most fundamental kind. If you are walking in the woods and you see something that your peripheral vision registers as snake-shaped and coiled, the appropriate response is to jump back first and look more carefully second. Not because snakes are always dangerous, but because the cost of responding to a harmless stick as though it were a snake is trivial compared to the cost of responding to a snake as though it were a harmless stick.

The neural mechanism that implements this rapid, automatic response was identified by the neuroscientist Joseph LeDoux, whose decades of research on the amygdala at New York University constitute one of the most important bodies of work in the history of emotional neuroscience. LeDoux discovered that the amygdala receives sensory information through two distinct pathways, which he memorably called the high road and the low road. The high road is the pathway through the cortex: sensory information from the thalamus is relayed to the appropriate sensory cortex, where it is processed in detail, identified with precision, and evaluated in the context of everything the brain knows about the current situation before a response is generated. This is the pathway of careful, accurate, nuanced perception: you see the thing, you analyze it fully, you decide whether it is a snake or a stick, and then you respond appropriately.

The low road is different. It is a direct pathway from the thalamus to the amygdala, bypassing the cortex entirely. This pathway is faster than the cortical route, cruder in its information processing, and capable of generating a fear response before the cortex has had any opportunity to evaluate the incoming information. It is the pathway that makes you jump before you think. The information arriving via the low road is not detailed enough to distinguish a snake from a stick with any reliability. But it arrives fast enough to initiate the fear response before the detailed analysis is complete, giving the body a head start on the preparation for potential danger.

This head start comes at a cost: false alarms. The low road, in its haste, generates fear responses to things that are not actually dangerous, including harmless sticks that look like snakes, and more generally, any stimulus that resembles the features of a past threat in some but not all respects. Most of the time, the cortex catches up within a few hundred milliseconds, evaluates the low-road alarm in the context of the full sensory picture, and either confirms or cancels the fear response. But there are circumstances, particularly when the fear response is very intense or when the cortical evaluation is impaired by fatigue, stress, or mental illness, where the low-road response runs ahead of cortical control and produces fear reactions that are disproportionate, persistent, or attached to stimuli that pose no real threat. This is the neural substrate of phobias, of panic attacks, and of some of the most characteristic symptoms of post-traumatic stress disorder.

The Chemistry of Love

In 2000, a team of researchers at University College London led by the neurologist Andreas Bartels and the neuroscientist Semir Zeki published a paper in NeuroReport that described something no one had ever previously attempted: an fMRI study of the neural correlates of romantic love. They recruited seventeen young adults who described themselves as truly, deeply, and madly in love with a romantic partner, people who thought about their partner many times a day, who felt a deep craving to be with them, who experienced the specific altered state of consciousness that the poets have been trying to describe for millennia. Each subject was shown photographs alternately of their beloved and of familiar friends of the same gender and age. The brain responses to the two categories of photographs were compared.

The results were striking enough to attract international attention. Viewing photographs of the beloved, compared to familiar friends, produced specific activation in four brain regions: the medial insula, the anterior cingulate cortex, and two subcortical areas whose involvement was particularly interesting given what scientists already knew about them, the caudate nucleus and the putamen. These subcortical regions are part of the basal ganglia, the brain's reward and habit system, and are rich in dopamine receptors. They are the same regions that are activated by other powerfully rewarding experiences:

by food when you are hungry, by water when you are thirsty, by music you love. And, as subsequent research confirmed, by cocaine.

This is not a metaphor. The neural circuitry of romantic love overlaps substantially with the circuitry of drug addiction. Both involve the dopaminergic reward system. Both produce a strong craving for the source of reward and a heightened response to cues associated with it. Both are characterized by narrowed attention, by the tendency to focus obsessively on the object of desire at the expense of other concerns. Both can produce withdrawal-like symptoms when the source of reward is removed. And both, in their acute phases, are accompanied by elevated levels of dopamine and norepinephrine in the brain, which produce the heightened arousal, the reduced need for sleep, the racing thoughts, and the extraordinary motivational energy that characterize early romantic love.

The analogy has limits, and it is important not to push it too far. Romantic love is not simply addiction, for reasons that go well beyond the fact that its social consequences are generally more positive. The brain regions activated by romantic love also include areas involved in empathy and in the theory of mind, the capacity to understand other people's mental states, that are not prominently involved in drug reward. And the longer-term phases of love, what is sometimes called companionate love as opposed to passionate love, are associated with different neural patterns that emphasize attachment, care, and social bonding rather than the acute dopaminergic craving of the early phase. But the overlap between love and reward, between the experience of caring deeply about another person and the brain systems that evolved to drive animals toward biologically essential resources, is real and it is informative: it tells us something important about why love is so powerful, so motivating, so central to human experience, and so utterly devastating when it is lost.

The Neuroscience of Grief

There is a prediction error at the heart of grief. Not a mistake, not a cognitive failure, but a prediction error in the technical sense that neuroscientists use the term: a mismatch between what the brain predicts and what it actually receives. The brain is, at one level, a sophisticated prediction machine. It builds models of the world, of the regularities and patterns and causal relationships that allow it to navigate experience efficiently, and it uses these models to generate predictions about what will happen next. When a prediction is correct, the brain notes the match and updates its model minimally. When a prediction is incorrect, the brain registers the mismatch as a signal that the model needs to be revised.

When someone you love dies, a person who has been woven into the predictive fabric of your brain for years or decades, every moment of the day that would normally generate a prediction about that person, an expectation of their presence, a calculation of what they might be doing right now, a habit of considering what they would think about this or that, now generates a prediction error. The model says they are there. Reality says they are not. The mismatch must be registered, processed, and used to gradually update a model that is, after a long relationship, deeply and extensively built around the assumption of that person's continued existence.

This is the neural account of grief, and it is a powerful one, not because it replaces the emotional experience of grief with a cold computational description but because it explains something that grieving people often report: the way that grief seems to have triggers everywhere, to be ambushed by unexpected reminders, to return with full force when a familiar habit encounters the absence that should not be there. Every trigger is a moment when the brain's ongoing prediction machinery generates an expectation that is then contradicted by reality, and every contradiction requires renewed processing of the mismatch, renewed integration of the loss into the model.

Neuroimaging studies of bereaved individuals have confirmed that the brain's activity during grief involves many of the same systems involved in physical pain, in reward processing, and in social cognition. The anterior cingulate cortex, a region involved in the experience of both physical and social pain, is active in the grief response. The nucleus accumbens, a key node in the reward system, shows activity that some researchers interpret as the brain's continued orientation toward the lost person as a source of reward, a kind of neural searching that underlies the common grief experience of an automatic movement toward wanting to share something with the person before remembering that they are gone. And the hippocampus, the memory system, is constantly engaged in grief, processing the flood of autobiographical memories that loss precipitates.

The Regulatory System

The picture of emotion that the neuroscience presents is not one of the brain as a purely reactive system, at the mercy of its own emotional machinery. There is a regulatory counterpart to every emotional generation system: a set of structures and circuits, primarily in the prefrontal cortex, that modulate, suppress, redirect, and contextualize emotional responses in the light of goals, values, social norms, and the demands of the current situation.

The prefrontal cortex exerts its regulatory influence over emotion primarily through descending connections to the amygdala, the insula, and the other limbic structures that generate emotional responses. When you choose not to express anger in a professional context, when you redirect your fear response by reappraising a situation as less threatening than it initially appeared, when you suppress your grief for the duration of a difficult meeting and allow yourself to feel it later in private, you are exercising prefrontal regulatory control over limbic emotion generation. This control is not effortless. It consumes cognitive resources, and it can be degraded by fatigue, stress, cognitive load, and certain mental health conditions.

The relationship between the prefrontal cortex and the amygdala is a central theme in the neuroscience of emotional regulation, and it is also a central theme in the neural story of mental illness. Depression, anxiety, PTSD, and borderline personality disorder all involve, to varying degrees, a disruption of the balance between emotional generation and emotional regulation: either the generating systems are overactive, or the regulatory systems are underactive, or, most commonly, both. Understanding these conditions as disorders of prefrontal-limbic balance, rather than as disorders of moral weakness or emotional instability, is one of the most important contributions that neuroscience has made to the reduction of stigma around mental illness.

The electrical life of your emotions is not a simple story. It is not reducible to a single brain region or a single neurotransmitter or a single circuit. It is a story of interaction and balance, of generating systems and regulating systems, of ancient subcortical structures and recently evolved cortical ones, of rapid automatic responses and slower, more deliberate evaluations. But it is a story that neuroscience is now telling with increasing precision and detail, and it is a story that, even in its current incomplete form, reveals something genuinely important: that the emotions that feel so fundamentally personal, so irreducibly subjective, so far beyond the reach of biology, are in fact among the most exquisitely biological things about you. They are what a brain does when it needs to navigate a social world. And they are, like everything else the brain does, electricity.

CHAPTER SIX

The Sleeping Brain

Why Your Brain's Nightly Electrical Storm Is the Most Important Thing It Does

IT IS ELEVEN THIRTY-seven in the evening. A man we will call David has been lying in bed for twelve minutes, and the transition is already beginning. His eyes are closed. His breathing, which was a few minutes ago the deliberate, slightly effortful breathing of a person trying to sleep, has softened into something more automatic, more unconscious, the rhythm of a body beginning to hand itself over to processes it cannot consciously direct. In the electroencephalography laboratory where he is spending tonight as a research volunteer, the computer screen monitoring his brain activity shows the shift: the alpha waves that characterized his relaxed wakefulness, those smooth, synchronized ten-cycle-per-second oscillations, are becoming less regular, interrupted increasingly by slower, higher-amplitude waves, and the technician watching the screen makes a note. Stage One. The descent has begun.

What is about to happen in David's brain over the next eight hours is not rest, in any meaningful sense of that word. It is work of a different kind, as metabolically demanding in its own way as anything his brain does during wakefulness, and more essential to his survival than almost anything else he will do today. The popular image of sleep as a passive state, a nightly switching-off of the brain's activity in order to recover from the demands of wakefulness, is not merely incomplete. It is almost precisely wrong. The sleeping brain is an extraordinarily active brain, cycling through states of electrical activity as diverse and as functionally distinct as any that occur during the waking day, performing operations of memory consolidation, emotional processing, toxin clearance, and neural maintenance that cannot be adequately performed while the organism is awake. To understand what sleep is, you must first abandon the idea that it is an absence. It is a presence: a specific, elaborate, precisely organized form of neural activity that your brain absolutely requires, and that the history of sleep deprivation research has shown, with brutal clarity, is as essential to survival as food or water.

The Architecture of the Night

Sleep is not a single state. It is a succession of states, organized into a repeating cycle that occurs four to six times over the course of a full night's sleep, each cycle lasting approximately ninety minutes in adults. Within each cycle, the brain moves through a sequence of distinct stages, each with its own characteristic electrical signature, its own pattern of brain wave activity, its own specific physiological and psychological character. Understanding this architecture is the foundation for understanding everything else about sleep: what it does, why it matters, and what happens when it goes wrong.

The sleep stages are divided into two major categories: non-rapid eye movement sleep, or NREM, and rapid eye movement sleep, or REM. NREM sleep is itself subdivided into three stages of increasing depth. Stage One NREM, the transitional stage in which David currently finds himself, is the lightest stage of sleep: the stage from which a person can be easily awakened, in which the boundary between wakefulness and sleep is not yet firmly established, and in which the hypnagogic hallucinations that some people experience at sleep onset, the sudden vivid images or sounds that jerk one back to wakefulness with a start, are most likely to occur. The brain waves of Stage One are a mixture of alpha activity and slower theta waves, and the eyes, though closed, may drift slowly from side to side in the rolling, unfocused movements that EEG technicians use to mark the stage transition.

Stage Two NREM is the first stage of genuine, consolidated sleep. The eyes are still. The body temperature has begun to drop. The heart rate and breathing have slowed further. And the EEG reveals two phenomena that are the hallmarks of Stage Two: sleep spindles and K-complexes. Sleep spindles are brief bursts of rapid, synchronized electrical activity, lasting between half a second and two seconds and occurring at a frequency of twelve to fifteen hertz, generated by the interaction between the thalamus and the cortex. They appear on the EEG as brief, waxing-and-waning oscillations superimposed on the slower background rhythm, and their function, which researchers have spent decades trying to fully understand, appears to be related to the active suppression of incoming sensory information, the brain's way of insulating the sleeping mind from the disturbances of the external environment. People who generate more sleep spindles are generally better at sleeping through noise and other disruptions than those who generate fewer.

K-complexes are large, dramatic electrical events: a sharp negative wave followed by a positive wave that produces the largest-amplitude electrical event that can be reliably recorded by a scalp EEG under normal circumstances. They can be triggered by external stimuli, a sudden noise or a touch, suggesting that one of their functions may be related to the evaluation of environmental threats during sleep, a kind of neural sentinel that monitors the environment for danger without fully waking the sleeper. But K-complexes also occur spontaneously, without any identifiable external trigger, and their full range of functions remains incompletely understood.

Stage Three NREM, often called slow-wave sleep or deep sleep, is the most restorative stage and the one most closely associated with the physiological functions that make sleep essential. The brain waves of slow-wave sleep are the delta waves: slow, large-amplitude oscillations of one to four hertz that represent the most synchronized electrical activity the waking or sleeping brain ever generates. During these slow waves, millions of neurons are firing together in coordinated bursts and then going collectively silent, producing the characteristic up-and-down pattern that gives delta sleep its distinctive EEG signature. It is very difficult to wake someone from slow-wave sleep, and if awakened they typically feel deeply disoriented, confused, and unrefreshed, a state that sleep scientists call sleep inertia and that reflects the disruption of a process that was not ready to be interrupted.

What Deep Sleep Actually Does

The most important function of slow-wave sleep, and the one that has attracted the most research attention in the past decade, is the clearance of metabolic waste from the brain. The discovery of the glymphatic system by Maiken Nedergaard and her colleagues at the University of Rochester, published in Science in 2013, was one of the most significant findings in neuroscience of the early twenty-first century, and it fundamentally changed the scientific understanding of why sleep is necessary.

The brain is, as noted in Chapter Two, an extraordinarily metabolically active organ. Its intensive activity generates waste products that must be cleared, just as intensive activity in any other tissue generates waste that must be removed. For most of the body, this waste clearance is handled by the lymphatic system. But the brain is separated from the rest of the body by the blood-brain barrier, and for many decades neuroscientists were puzzled by the apparent absence of a brain-specific waste clearance mechanism. The glymphatic system resolved this puzzle. It consists of a network of channels surrounding the brain's blood vessels, lined with specialized cells called astrocytes, through which cerebrospinal fluid is pumped during sleep to flush out the waste products that have accumulated in the brain tissue during waking hours.

During wakefulness, the channels of the glymphatic system are relatively narrow, and the flow of cerebrospinal fluid through the brain tissue is limited. During slow-wave sleep, these channels expand dramatically, increasing in diameter by approximately sixty percent, and the flow of cerebrospinal fluid through the brain increases correspondingly. This nightly flushing removes a wide range of metabolic waste products, but the one that has attracted the most attention is amyloid-beta, the protein that forms the plaques characteristic of Alzheimer's disease. During wakefulness, amyloid-beta accumulates in the spaces between brain cells. During sleep, the glymphatic system clears it away. Sleep deprivation, even a single night, leads to a measurable accumulation of amyloid-beta in the brain that is not fully resolved by subsequent sleep. Chronic sleep deprivation, maintained over years or decades, may contribute to the amyloid accumulation that is one of the hallmarks of Alzheimer's pathology.

This is not a speculative hypothesis. It is supported by multiple independent lines of evidence, including studies using positron emission tomography to measure amyloid burden in the brains of sleep-deprived human volunteers, longitudinal studies associating poor sleep quality in midlife with increased Alzheimer's risk decades later, and animal studies in which disruption of glymphatic function accelerates amyloid accumulation and cognitive decline. The implication is clear and important: adequate sleep is not a lifestyle luxury. It is a neuroprotective necessity, one of the most powerful tools available for reducing the risk of neurodegenerative disease.

Slow-wave sleep also plays a central role in memory consolidation. During the slow oscillations of deep sleep, the hippocampus and the neocortex engage in a coordinated electrical dialogue in which the memories encoded in the hippocampus during the previous waking period are replayed and transferred to long-term storage in the neocortex. This replay occurs at compressed timescales: the neural patterns that represented a specific experience during waking are reactivated during sleep in a rapid, sequential fashion, strengthening the synaptic connections that encode the memory and facilitating its gradual independence from the hippocampus. Disrupting slow-wave sleep, either by sleep deprivation or by selective suppression of delta waves, impairs this consolidation process and reduces the retention of newly learned information.

Into the Dream

After approximately ninety minutes of NREM sleep, cycling from Stage One through Stage Two and Stage Three and back again, David's brain makes a dramatic transition. The slow, synchronized delta waves of deep sleep give way to a pattern that, if you were watching the EEG without knowing what you were seeing, you might mistake for wakefulness: rapid, desynchronized, low-amplitude activity that looks almost identical to the brain waves of an alert, engaged mind. David's eyes begin to move, rapidly, in coordinated bursts, beneath his closed lids. His body becomes paralyzed, the large muscles of his limbs and trunk held in a state of active suppression by signals from the brainstem that prevent him from acting out what his brain is about to generate. His heart rate becomes irregular. His breathing deepens and quickens. He is entering REM sleep. He is beginning to dream.

REM sleep is, in many respects, the most extraordinary of the brain's states. The EEG pattern is virtually indistinguishable from wakefulness, yet the person is deeply unconscious by behavioral criteria, unable to respond to their environment and very difficult to wake. The brain is generating vivid, narratively coherent, emotionally intense experiences in the complete absence of sensory input, conjuring them from internal electrical activity with a richness and a vividness that rivals, and sometimes surpasses, the texture of waking perception. The body is paralyzed to prevent the dreaming mind from acting out its generated scenarios, a paralysis so complete that it can sometimes persist briefly after waking, producing the terrifying experience of sleep paralysis in which the person is conscious but unable to move, often accompanied by vivid hallucinations of a threatening presence in the room.

What is dreaming for? This is one of the most debated questions in sleep science, and honest scholars will tell you that it does not yet have a definitive answer. Several compelling hypotheses have accumulated evidence in their favor, and they are not mutually exclusive. One influential proposal, developed by the neuroscientist Matthew Walker at the University of California at Berkeley, is that REM sleep serves as a kind of overnight emotional therapy: a period during which the brain reprocesses emotionally charged experiences from the recent past, integrating them into the larger fabric of autobiographical memory while somehow stripping them of their acute emotional charge. The evidence for this comes partly from the finding that depressed patients who are prevented from entering REM sleep actually show improvements in their mood, while those who are allowed to sleep normally do not, suggesting that REM sleep in depression may be replaying negative emotional material in a way that maintains or deepens the depressive state rather than resolving it.

Another influential hypothesis, supported by a rich body of animal and human research, is that REM sleep supports a particular kind of memory consolidation that is distinct from the declarative memory consolidation of slow-wave sleep: the consolidation of procedural and emotional memories, the integration of new learning with existing knowledge frameworks, and the discovery of hidden patterns and relationships in recently acquired information. The experience, reported by many scientists, artists, and problem-solvers, of waking from sleep with a solution to a problem that had resisted conscious analysis the previous evening, reflects this integrative function of REM sleep: the dreaming brain is not merely replaying past experience but actively combining and recombining elements of it in ways that can produce novel insights.

The Cost of Insufficient Sleep

The science of sleep deprivation is, in some respects, the most alarming body of research in all of neuroscience, because its findings are so clear, so consistent, and so directly relevant to the lives of so many people in the modern world. The industrialized world is in the grip of what the sleep scientist Matthew Walker has called a global sleep deprivation epidemic, with studies consistently finding that a large fraction of adults in wealthy, developed nations regularly sleep less than the seven to nine hours that the evidence indicates is necessary for optimal health. The consequences of this chronic sleep deficit are documented with the same kind of large-scale, reproducible evidence that we associate with the strongest findings in epidemiology and pharmacology.

A single night of sleeping only four to six hours, compared to a full eight-hour sleep opportunity, produces measurable impairments in attention, working memory, emotional regulation, and immune function. Reaction time slows. Error rates increase. The prefrontal cortex, which regulates emotional responses and enables flexible, contextually appropriate behavior, shows reduced activity in neuroimaging studies of sleep-deprived subjects. The amygdala, which generates emotional responses and evaluates threats, shows increased reactivity, and the regulatory connection between the prefrontal cortex and the amygdala is weakened. The sleep-deprived brain is effectively a more emotionally reactive and less rationally regulated brain, one that is more likely to respond to provocation with anger and less able to inhibit inappropriate behavioral impulses.

The cardiovascular consequences of chronic sleep deprivation are equally well documented. Short sleepers, defined in most studies as people who regularly sleep less than six hours per night, have significantly higher rates of hypertension, coronary artery disease, heart failure, and stroke than those who sleep seven to nine hours. The mechanisms are multiple: sleep deprivation elevates cortisol and sympathetic nervous system activity, raising blood pressure and heart rate. It increases circulating levels of inflammatory markers. It impairs the regulation of blood glucose, increasing the risk of type 2 diabetes. And it disrupts the normal nocturnal reduction in blood pressure that allows the cardiovascular system to recover from the demands of the waking day.

The immune consequences are equally striking. Studies in which subjects were deliberately exposed to rhinovirus, the common cold virus, after having their sleep monitored for several weeks found that those who averaged less than seven hours of sleep per night were almost three times as likely to develop a cold as those who averaged eight hours or more, independent of age, stress level, smoking habits, and other factors. Vaccination studies have found that sleep-deprived subjects generate significantly smaller antibody responses to flu vaccines than well-slept subjects, suggesting that the immune benefit of vaccination is substantially reduced in people who are not sleeping adequately.

The relationship between sleep and mental health is bidirectional and particularly consequential. Poor sleep is both a symptom and a cause of depression, anxiety, bipolar disorder, and schizophrenia. Sleep disturbance is one of the most reliable predictors of psychiatric relapse in people with established mental health conditions. And in healthy individuals, experimental sleep deprivation produces measurable increases in anxiety and in the emotional lability that characterizes the early stages of mood disorders. The causal arrow runs in both directions: mental health problems disrupt sleep, and disrupted sleep worsens mental health problems, creating vicious cycles that can be very difficult to break without addressing both sides simultaneously.

The Night's Final Gift

By six in the morning, David is in his fourth REM period of the night, the longest and most vivid of the night's dreams. His brain is doing something that, if described to him in waking terms, might sound like a kind of biological alchemy: it is taking the experiences of the previous day, the facts he studied, the conversations he had, the anxieties he felt, and it is transforming them. Not erasing them. Not simply filing them away. Transforming them, integrating them with the deeper structures of his knowledge and experience, finding patterns he did not consciously notice, forging connections he was not deliberately making, processing the emotional residue of experiences that were too complex or too charged to be fully digested in the moment they occurred.

When he wakes, in twenty minutes, he will not remember the dream. He almost certainly will not know that this transformation happened. He will simply feel, if the night has gone well, the particular clarity and readiness of a brain that has done its nightly work properly: alert, emotionally balanced, able to focus and sustain attention and regulate his responses to the inevitable provocations of the day ahead. He will feel, in the word that seems most accurate to describe it, himself. That quality of selfhood, that readiness for the day, is not something that happens despite the night. It is something that the night produced. The sleeping brain, in all its electrical complexity, gave it to him. And tomorrow night, if he allows it, it will give it to him again.

CHAPTER SEVEN

The Hardest Problem

Consciousness, the Self, and Why Science Has Not Yet Solved Its Greatest Mystery

"Why is there something it is like to see the color red, to feel pain, to experience joy? Why are not all these functions performed in the dark, without any inner light of experience?"

— David Chalmers, The Conscious Mind, 1996

I WANT TO BEGIN THIS chapter with a confession. It is a confession that I suspect many scientists who work on consciousness would make if they felt entirely free to be honest in public, which the culture of scientific credentialing does not always encourage. The confession is this: I do not know what consciousness is. Neither does anyone else. Not definitively. Not in the way that we know what an action potential is, or what a memory is at the synaptic level, or what the NMDA receptor does when it opens its ion channel. There are theories of consciousness, sophisticated and carefully developed theories, some of which we will examine in this chapter. There is a large and growing literature of empirical research on the neural correlates of consciousness, the brain activities that accompany and in some sense support conscious experience. But there is no consensus theory, no agreed mechanism, no widely accepted explanation of how and why the physical activity of neurons gives rise to the subjective, first-person, irreducibly personal experience of being aware.

This is unusual for a science chapter. Scientists are not typically in the business of beginning their discussions with admissions of fundamental ignorance. But the Hard Problem of consciousness, a term coined by the philosopher David Chalmers in 1995 and now used universally in the field, is unusual. It is a problem that has resisted every approach that has worked for every other problem in the history of neuroscience. And understanding why it is hard, really understanding the source of the difficulty, is itself one of the most illuminating things you can do if you want to understand the brain.

Easy Problems and the Hard Problem

David Chalmers drew a distinction that has organized the philosophy of mind ever since between what he called the easy problems of consciousness and the Hard Problem. The easy problems, he was careful to note, are not actually easy in any ordinary sense of the word. Understanding how the brain processes visual information, how it integrates signals from different sensory modalities, how it directs attention selectively to some stimuli and away from others, how it enables the reporting of mental states, how it distinguishes waking from sleeping and controls the sleep-wake transition: these are all genuinely difficult scientific problems, requiring sophisticated experiments, complex mathematical models, and decades of painstaking research. But they are tractable in principle, because they are all questions about the mechanisms by which the brain processes information and produces behavior. They ask how the brain does specific functional things, and science has shown itself to be very good at answering how-does-it-work questions about biological mechanisms.

The Hard Problem is different in kind, not just in degree. It does not ask how the brain processes information about color. It asks why the processing of information about wavelengths of light in the range of 620 to 750 nanometres is accompanied by the experience of redness: a specific, irreducible, qualitative character of awareness that is not contained in any description of the wavelengths or the neurons or the processing. It does not ask how the brain detects tissue damage and generates a pain signal. It asks why the detection of tissue damage is accompanied by suffering: why there is something that it feels like to be in pain, rather than the pain being processed in the dark, automatically and without any inner experience at all.

The gap between the mechanism and the experience, between what the brain does and what it feels like to be a brain doing it, is what Chalmers called the explanatory gap. It is not a gap in our knowledge of the mechanisms. We know a great deal about the mechanisms, and our knowledge is improving rapidly. It is a gap between the mechanistic description, however complete it becomes, and the fact of experience. Even a complete, exhaustive description of everything that happens in the brain when a person sees red would not, in any obvious way, explain why they experience redness. There would always remain the further question: but why does all of this processing feel like something?

Integrated Information Theory

The most mathematically developed theory of consciousness currently in circulation is Integrated Information Theory, or IIT, developed by the Italian-American neuroscientist Giulio Tononi at the University of Wisconsin-Madison. Tononi's theory begins from phenomenology: from a careful analysis of the properties of conscious experience as it is actually lived. He identifies five axiomatic properties that all conscious experience must have: it exists intrinsically, meaning it does not depend on external observation; it is structured, with parts that are causally related to each other; it is specific, with a definite character that distinguishes it from other possible experiences; it is unified, forming a single whole rather than a collection of independent parts; and it is definite, excluding all the alternative experiences it is not having.

From these phenomenological axioms, Tononi derives a mathematical quantity he calls phi, which measures the degree to which a system integrates information: the extent to which the system as a whole generates more information than the sum of its parts. A system with high phi is one whose overall state cannot be reduced to the states of its individual components, because the connections between the components create information that only exists at the level of the whole system. Tononi's central claim, the part that makes IIT genuinely interesting as a theory of consciousness, is that consciousness is identical to integrated information: that the degree of consciousness in a system is proportional to its phi value, and that consciousness therefore exists wherever high phi exists, in humans and animals and possibly in other systems depending on their informational architecture.

IIT makes some predictions that match well with our intuitions about consciousness and some that are deeply counterintuitive. On the intuitive side, it predicts that the cerebral cortex, which is densely and richly interconnected in ways that generate very high phi, is the primary seat of conscious experience in the brain, while the cerebellum, which contains more neurons but is organized in a much more modular and less interconnected way that generates lower phi, contributes little to conscious experience. This is consistent with the clinical observation that patients who lose large portions of their cerebellum often show remarkably little impairment of consciousness, while patients who lose corresponding portions of their cortex typically experience dramatic changes in the quality of their conscious experience.

On the counterintuitive side, IIT predicts that any physical system with high integrated information is conscious, regardless of its substrate. A sufficiently interconnected network of simple logic gates could, in principle, be conscious if its phi value were high enough. A future computer designed with the right architecture would be conscious. And some very simple biological systems, if their informational architecture generates sufficient phi, might have a small but genuine measure of consciousness. This last prediction strikes many scientists and philosophers as absurd, and it has generated fierce criticism of IIT from people who feel that the theory is too permissive, that it attributes consciousness too liberally to systems that seem obviously not to have it. Tononi's response is that this reaction reflects our intuitive prejudices about what kinds of things can be conscious, prejudices that may be as unfounded as earlier intuitions about what kinds of things could have mathematical structure.

Global Workspace Theory

A very different account of consciousness is provided by Global Workspace Theory, developed by the American cognitive scientist Bernard Baars in the 1980s and extended by the French neuroscientist Stanislas Dehaene in the years since. Where IIT begins from phenomenology and derives neural implications, Global Workspace Theory begins from cognitive science and neuroscience and attempts to explain what consciousness does and how it arises from the architecture of the brain.

The central metaphor of Global Workspace Theory is a theater. In a theater, most of the activity takes place backstage, out of sight of the audience: the stage crew, the directors, the prompters, the technicians all do their work invisibly, supporting the performance without being part of it. What the audience sees is only what the spotlight illuminates on the stage. In Global Workspace Theory, the brain is the theater, the various specialized processing systems are the backstage workers, and consciousness is what happens when information is broadcast into the global workspace, a distributed network of long-range cortical connections that acts as the spotlight, making information widely available to multiple cognitive systems simultaneously.

The global workspace is not a specific brain region but a network: one that includes the prefrontal cortex, the parietal cortex, and the long-range connections that link them to each other and to sensory processing areas. Information that enters the global workspace, that is broadcast into consciousness, becomes available to all the systems connected to the network: memory systems, planning systems, language systems, motor control systems. This global availability is what makes information conscious in the Global Workspace account, and it is contrasted with the vast amount of information processing that occurs in the brain outside the workspace, in the specialized backstage systems that do their work automatically and unconsciously, without broadcasting their operations into the global network.

The neural evidence for Global Workspace Theory is substantial. Neuroimaging studies of the neural correlates of consciousness consistently find that conscious perception, compared to unconscious processing of the same stimulus, is associated with a widespread ignition of cortical activity that includes the frontal and parietal regions of the global workspace network, whereas unconscious processing activates only the local cortical regions specialized for that type of stimulus. Studies of the timing of neural responses to conscious versus unconscious stimuli find that conscious perception involves a late-occurring, large-amplitude wave of cortical activity approximately 300 to 500 milliseconds after stimulus onset, far later than the early sensory responses that occur for both conscious and unconscious stimuli. And studies of patients with disorders of consciousness, including vegetative state and minimally conscious state, find that the ability to generate the global ignition response to stimuli is one of the most reliable markers of preserved conscious awareness.

The Neural Correlates of Consciousness

Whatever theory of consciousness ultimately proves most correct or most useful, the empirical search for what neuroscientists call the neural correlates of consciousness, the specific brain activities that accompany conscious experience, has produced a body of findings that are interesting and important regardless of their theoretical interpretation. These findings tell us something real about the brain basis of awareness, even if they do not fully explain it.

Perhaps the most robust finding in the neuroscience of consciousness is the role of the posterior hot zone, a collection of cortical regions at the back of the brain, including the posterior parietal cortex, the posterior temporal cortex, and the occipital cortex, that appears to be particularly important for the specific content of conscious experience: what you are aware of, what your experience consists of at any given moment. Damage to or disruption of these posterior regions produces specific alterations in the content of consciousness: lesions in the visual cortex produce specific visual deficits, lesions in the somatosensory cortex produce specific tactile deficits, lesions in the posterior parietal cortex can produce dramatic disorders of spatial awareness including the fascinating syndrome of hemispatial neglect in which patients entirely fail to attend to one side of their visual field and even one side of their own body.

The frontal cortex, by contrast, appears to be important for the reporting of conscious experience and for higher-order cognitive operations performed on conscious content, rather than for the content itself. Patients with frontal lobe damage often show striking preserved conscious perception but impaired ability to report, reflect on, or act on what they perceive. This dissociation between the content of consciousness and the reflective, reportable aspect of consciousness is one of the most theoretically interesting findings in the field, and it maps onto a distinction that some philosophers have drawn between phenomenal consciousness, the raw experience of seeing red or feeling pain, and access consciousness, the availability of information for use in reasoning, reporting, and behavioral control.

The View from the Inside

All of the theories and findings described above have something in common: they are third-person accounts of consciousness, descriptions of it from the outside, in the language of mechanisms and correlates and information processing and neural activity. They are, as theories of physical mechanisms always are, descriptions of what can be observed and measured by an external observer. And the Hard Problem is precisely the problem that no third-person account, however complete, seems sufficient to explain the first-person fact of experience: the fact that there is something it is like to be you.

This is not a problem that can be solved by getting more data or building better brain scanners. It is a conceptual problem, a problem about the relationship between two very different kinds of description: the objective, third-person description of physical processes and the subjective, first-person description of experience. Whether these two kinds of description are ultimately reconcilable, whether the gap between them can be bridged by a sufficiently sophisticated theory of information or causation or integrated information, or whether it represents a fundamental feature of reality that cannot be explained in purely physical terms, is perhaps the deepest open question in all of human knowledge.

There are scientists and philosophers who believe that the Hard Problem will eventually be solved by a sufficiently advanced neuroscience, that consciousness will turn out to be something that a complete physical description of the brain entails, even if we cannot yet see how. There are others who believe that the Hard Problem reveals something fundamental about the limits of physical explanation, that consciousness is either a basic feature of reality that cannot be explained in terms of anything simpler, as IIT in some interpretations suggests, or that it is simply beyond the reach of any third-person science. And there are still others who believe that the Hard Problem is, in some sense, a philosophical confusion: that once we have explained all the functional mechanisms of the brain, there will be nothing left to explain, and that the apparent gap between mechanism and experience will dissolve.

I do not know which of these positions is correct. I am not sure anyone does. What I am sure of is that the question matters: that understanding the relationship between the physical and the experiential is not a luxury of philosophy but a necessity if we are to understand what we are, what our machines might become, and what moral status we should extend to the many kinds of minds that exist or may come to exist in the world. These are questions we will return to in the book's final chapters, where the technology of reading and writing the brain's electricity has made them not merely philosophical but urgently practical. For now, let us hold the mystery. The light shines. We do not yet know why.

PART THREE

WHEN THE ELECTRICITY GOES WRONG

Brain Disorders Reframed as Diseases of Electrical Signalling

Part Introduction

The Misfiring Machine

There is a metaphor that has dominated public understanding of mental illness and neurological disease for most of the twentieth century, and it has caused an enormous amount of unnecessary suffering. The metaphor is moral. In this metaphor, the person with depression is not sufficiently resilient or grateful. The person with schizophrenia has lost touch with reality in a way that implies some failure of their rational agency. The person with epilepsy is, in some cultures and in some historical periods, possessed by spirits, or punished by gods, or simply different from normal people in ways that make them objects of fear and stigma rather than care and respect. The person with Alzheimer's disease is gradually becoming less of a person, their identity dissolving with their memories, until the body that remains is just an echo of the one who lived in it.

The neuroscience of the past half-century has provided the evidence to dismantle this moral metaphor, though the dismantling is still very much in progress. What the evidence shows, with clarity that only improves as the science advances, is that depression is a disorder of specific brain circuits, involving abnormal patterns of electrical activity in identifiable neural networks, as physical in its nature as a cardiac arrhythmia or a broken bone. Epilepsy is a disorder of electrical synchronization, a short circuit in the brain's electrochemical machinery, no more a reflection of supernatural influence than a faulty electrical connection in a building's wiring system. Schizophrenia involves measurable abnormalities in the timing and coordination of neural oscillations, particularly in the gamma frequency band, that produce the characteristic distortions of perception and reasoning not because the person has failed to think correctly but because the machinery of their thinking has been disrupted by changes, they did not choose and cannot easily reverse. Alzheimer's disease is a progressive destruction of the synaptic connections that encode identity and memory, driven by specific protein chemistry and neuroinflammation and electrical dysfunction, not a dissolution of the person but a destruction of the physical substrate that supported them.

Part Three of this book examines these conditions through the electrical lens that neuroscience has provided. It does not do so to reduce them to mechanisms or to suggest that the human experience of living with these conditions can be fully captured by a description of neurons and circuits and oscillations. It does so because understanding the electrical reality of these conditions is the foundation for developing better treatments, for reducing the stigma that still surrounds them, and for appreciating the extraordinary courage of the people who live with them and the extraordinary dedication of the scientists and clinicians who have spent careers trying to understand and relieve them.

CHAPTER EIGHT

The Storm Inside

Epilepsy, the Brain's Electrical Catastrophe, and What It Reveals About the Mind

SARAH IS STANDING IN the kitchen, pouring orange juice into a glass, when the world changes. It does not change dramatically, not at first. There is a sensation she has learned to recognize, a smell that is not there, like something burning though nothing is burning, and a feeling she can only describe as familiarity that does not belong, a sense that this moment has happened before in a way that is different from ordinary deja vu, more insistent, more total, more accompanied by a slow, rising anxiety that she is not generating so much as receiving. She sets the glass down on the counter. She has two, maybe three seconds to get to the chair.

She makes it to the chair. And then she is absent. Her eyes are open but unseeing. Her right hand moves in a repetitive, picking motion against the fabric of her shirt. She does not respond when her daughter calls her name from the other room. For approximately ninety seconds, the electrical storm sweeps through the temporal lobe of her brain, generating a complex experience that she will later describe in fragmented terms: images without context, emotions without cause, voices from somewhere that spoke words she could not quite hear. And then it passes. The abnormal electrical synchronization dissolves. The storm subsides. She comes back to herself slowly, confused, exhausted, not quite sure how much time has passed or what just happened, reaching for the glass she set down what feels like moments ago but was, her daughter tells her, almost two minutes.

Sarah has temporal lobe epilepsy. She has had it since she was twelve years old. She manages it with medication that reduces but does not eliminate her seizures. She is a teacher. She drives, on days when she is confident the medication is doing its work. She raises her daughter. She lives a full life, with the particular vigilance and the particular relationship with uncertainty that characterizes the lives of the forty-five to fifty million people worldwide who share her condition. Her epilepsy is, at one level, a neurological condition with a specific EEG signature, a specific anatomical origin, a specific set of risk factors and pharmacological targets. At another level, it is the central organizing fact of a particular life, shaping how she relates to her body, to uncertainty, to the moments between one experience and the next.

Both of these things are true simultaneously, and holding both of them in mind is what this chapter attempts.

The Oldest Brain Disorder

Epilepsy has been documented in human beings for at least four thousand years, making it the oldest recorded brain disorder in history. The Edwin Smith Surgical Papyrus, dating to approximately 1700 BCE and believed to be a copy of an even older Egyptian text, describes seizures with clinical accuracy that would not be out of place in a modern neurology textbook. The Code of Hammurabi, the Babylonian legal text from approximately the same period, addresses epilepsy in its provisions for slaves, noting that a slave who developed the condition after purchase could be returned for a refund, reflecting both the condition's recognizability and its social consequences. Ancient Sanskrit texts from the Vedic tradition describe seizures under the name apasmara, meaning loss of memory, with detailed observations about the different forms the condition could take and various treatments, largely ineffective by modern standards, for addressing it.

For most of this long history, epilepsy was interpreted through supernatural frameworks. In ancient Mesopotamia, it was attributed to demonic possession, and its treatment was primarily the province of exorcists and priests. In ancient Greece, it was called the sacred disease, a name that Hippocrates objected to strenuously in a treatise written around 400 BCE, arguing with remarkable prescience that epilepsy was a disorder of the brain and nothing more, and that attributing it to divine or demonic causes was not piety but ignorance. The Hippocratic text On the Sacred Disease is one of the earliest examples in recorded history of a naturalistic, brain-centered account of a neurological condition, and its central argument, that what appears to be sacred or supernatural in epilepsy is in fact the consequence of a physical disorder of the brain, remains as relevant and as contested in practice as it was when it was written 2,400 years ago.

The supernatural interpretation persisted throughout the medieval period in Europe and in many other cultural traditions. Epilepsy was associated with demonic possession, with lunar influence, with divine punishment, with moral failing. The treatments prescribed reflected these interpretations: prayer, exorcism, amulets, herbs associated with sacred power, and in some periods and places, social exclusion and institutionalization. It was not until the nineteenth century, with the development of clinical neurology as a medical discipline and the work of figures like the British neurologist John Hughlings Jackson, that a genuinely medical understanding of epilepsy began to emerge.

John Hughlings Jackson and the Electrical Understanding

John Hughlings Jackson is not a household name, but among neurologists he occupies a position of almost legendary importance as the person who, more than any other single individual, established the conceptual framework for understanding epilepsy as a disorder of the brain's electrical activity. Jackson practiced at the National Hospital for the Paralysed and Epileptic in London, which was founded in 1860 and remains operational today as the National Hospital for Neurology and Neurosurgery, the oldest and largest specialist neurology hospital in the world. He spent four decades there observing epileptic patients with the systematic thoroughness of a natural historian cataloguing a new species.

Jackson's central insight, developed in a series of papers published between the 1860s and the 1880s, was that the specific form a seizure took, the particular muscles that contracted, the particular sensations that were experienced, the particular behavioral changes that accompanied it, reflected the specific location in the brain from which the abnormal electrical discharge originated. He had observed that many of his patients had seizures that began with a specific, reproducible movement or sensation in one part of the body, and then spread in a predictable sequence to adjacent body parts, tracing what appeared to be a map of the motor cortex in reverse. He called these focal seizures, and the characteristic spreading pattern that he described became known as the Jacksonian march.

The significance of Jackson's observations extended far beyond epilepsy. By showing that specific motor and sensory functions could be systematically disrupted by focal electrical disturbances originating in specific brain regions, he provided compelling evidence for the functional localization of the brain: the idea that different regions of the cortex are specialized for different functions, and that a detailed map of those functions could in principle be constructed. Wilder Penfield, whose electrical stimulation maps of the human brain we encountered in Chapter Three, was building directly on the conceptual foundations that Jackson laid, using the controlled electrical stimulation of the open brain to do what Jackson had done by observing the uncontrolled electrical storms of epileptic seizures.

The Electrical Storm: What Actually Happens

A seizure is, at its most fundamental level, an abnormal and excessive synchronization of electrical activity in a population of neurons. Under normal circumstances, the brain's electrical activity is characterized by a certain degree of organized desynchronization: different neuronal populations are active at different times, in patterns that reflect the specific computations being performed, the specific information being processed, the specific behavioral demands being met. This desynchronization is essential for the brain's computational function. If all the neurons in a region fired at the same time, they would generate useless noise rather than useful information.

In epilepsy, the mechanisms that normally prevent excessive synchronization fail. The result is a wave of synchronized firing that sweeps through a neuronal population like a chain reaction, with each neuron's firing driving its neighbors to fire, until the abnormal synchrony has recruited a large population of cells into a self-sustaining electrical storm. The EEG signature of a seizure, the characteristic high-amplitude, highly synchronized rhythmic discharges that appear against the background of normal brain activity, is one of the most visually dramatic events in all of clinical neurophysiology, a signal so distinctive and so unmistakable that trained EEG readers can identify it instantly.

The particular character of a seizure, what the person experiencing it perceives and how their behavior changes, depends critically on where in the brain the storm originates and how far it spreads. A seizure that begins in the primary motor cortex and remains confined to a small area may produce only a brief, rhythmic jerking of one hand or one side of the face, without any loss of consciousness and without any subjective experience beyond the involuntary movement itself. A seizure that begins in the temporal lobe, as in Sarah's case, may produce the characteristic aura of temporal lobe epilepsy: the olfactory hallucinations, the intense deja vu, the autonomic changes, the altered emotional states, and the complex automatic behaviors that reflect the disruption of temporal lobe functions including memory, emotion, and complex sensory processing. And a seizure that rapidly spreads from its point of origin to involve both hemispheres of the brain, recruiting the reticular activating system in the brainstem that maintains wakefulness, produces loss of consciousness and the generalized convulsive seizure that most people recognize as the dramatic expression of epilepsy.

What Epilepsy Has Taught Us

Epilepsy has been, paradoxically, one of the most productive sources of knowledge about normal brain function that neuroscience has ever had. The focal nature of many seizures, combined with the possibility of studying the brain during seizures using intracranial EEG recording in patients undergoing surgical evaluation, has provided researchers with an extraordinary window into the relationship between specific brain regions and specific functions.

The recordings made by Wilder Penfield during awake surgical procedures, described in Chapter Three, were conducted in epilepsy patients whose brains needed to be mapped before surgical resection. The memory experiments of Itzhak Fried at the University of California at Los Angeles, who has implanted depth electrodes in the brains of epilepsy patients and recorded from individual neurons while subjects perform memory tasks, have produced some of the most detailed single-neuron recordings of human memory processes ever obtained. The study of ictal phenomena, the experiences that occur during seizures, has illuminated the neural basis of perception, memory, emotion, and the sense of self in ways that studies of the normal, seizure-free brain cannot easily replicate.

Perhaps most importantly, epilepsy has provided the clearest evidence in all of neuroscience for the functional localization of the brain: the demonstration that specific regions of the cortex are specialized for specific functions, and that disrupting those regions with either a seizure discharge or a surgical lesion produces specific and predictable deficits. The systematic relationship between seizure semiology, the particular form a seizure takes, and the location of its origin is so reliable that experienced epileptologists can often localize the seizure focus from a careful description of the patient's seizure phenomenology, before seeing any brain images or EEG recordings. The seizure, in this sense, is itself a diagnostic instrument, a forced experiment that reveals the function of the brain region it disrupts by temporarily destroying that function.

Treatment, Hope, and Limits

Approximately sixty to seventy percent of people with epilepsy can achieve adequate seizure control with antiepileptic medication. The drugs work through various mechanisms: some increase the brain's inhibitory tone by enhancing the activity of GABA, the brain's primary inhibitory neurotransmitter. Others reduce the activity of the excitatory neurotransmitter glutamate. Others stabilize the voltage-gated sodium channels whose opening initiates the action potential, making neurons less likely to fire the rapid, repetitive bursts that initiate seizure discharges. The choice of medication depends on the seizure type, the epilepsy syndrome, the patient's age, sex, other medical conditions, and individual tolerability.

For the thirty to forty percent of patients whose seizures are not adequately controlled by medication, the options are more limited and more invasive. Surgery to remove the seizure focus is the most effective treatment for drug-resistant epilepsy in carefully selected patients, producing seizure freedom in approximately sixty to seventy percent of those who undergo resection of a temporal lobe focus. Vagus nerve stimulation, in which an implanted device delivers regular electrical pulses to the vagus nerve, reduces seizure frequency in about half of patients who receive it, though rarely achieving complete seizure freedom. Responsive neurostimulation, in which electrodes implanted in the seizure focus detect the early electrical signatures of an impending seizure and deliver a brief counterstimulation to abort it before it becomes clinically manifest, is one of the most technically sophisticated current approaches and represents the growing edge of the field.

Sarah manages her epilepsy. She will always manage it. It is not a problem to be solved and put away but a condition to be lived with, a feature of her electrical brain that she has incorporated into the architecture of her life with a combination of medical compliance, personal courage, and the particular intimacy with her own neurology that only comes from years of learning to read the signals her brain sends before a storm. She knows her triggers: sleep deprivation, stress, the irregular rhythms of travel. She knows her warning signs. She knows the chair. And she knows, with the knowledge that only direct experience can produce, that her brain, for all its occasional electrical misbehavior, is also the organ that makes her a teacher, a mother, a person of intelligence and care and creativity and love. The storm is part of it. It does not define it.

CHAPTER NINE

The Dark Current

Depression, Anxiety, and the Broken Electrical Circuits of Mood

HE HAS BEEN ASKED TO describe it so many times, by so many doctors, in so many different clinical settings over fourteen years of treatment, that he has developed a repertoire of descriptions that he offers depending on how much time he thinks the doctor has and how much he thinks they will understand. For the five-minute appointments, he says: low mood, poor energy, difficulty concentrating, loss of interest in things I used to enjoy. For the doctors who seem to have more time and more patience, he says: it is like everything has lost its dimension. Like the world is a photograph of itself, flat, real in outline but missing the quality that makes it actually real, actually worth moving toward. He is forty-two years old. He has tried seven different antidepressants, three psychotherapists, one course of CBT, and a period of intensive exercise that helped for about three months before the flatness came back. He is not in crisis. He is not suicidal. He is simply living in a brain that has been doing something wrong, electrically and chemically wrong, for most of his adult life.

His name does not matter. What matters is that he is real, that he is one of more than 280 million people worldwide who currently live with major depressive disorder, and that the condition he is describing has a biological reality as specific and as measurable as any other disorder of a physical organ. It is not a failure of gratitude or perspective. It is not a consequence of insufficient effort or inadequate character. It is a disruption of specific brain circuits, involving specific patterns of abnormal electrical activity in identifiable neural networks, that is currently inadequately treated by the tools medicine has developed and that the emerging science of neuromodulation is beginning, haltingly and imperfectly but genuinely, to address.

The Circuit Disorder Model

The traditional neurochemical model of depression, the model that has dominated both research and clinical practice since the 1960s, proposes that depression results from a deficiency of certain neurotransmitters, particularly serotonin and norepinephrine, in the synaptic connections of specific brain circuits. This model was inferred, not directly from brain measurements, but from the observation that drugs that increased the availability of these neurotransmitters in the synapse, the monoamine oxidase inhibitors and the tricyclic antidepressants, produced improvements in depressive symptoms in many patients. The reasoning was straightforward: if drugs that increase serotonin help depression, then depression must involve too little serotonin.

The neurochemical model has been genuinely useful. The antidepressant medications it spawned, particularly the selective serotonin reuptake inhibitors that became available in the 1980s and that now include among the most widely prescribed drugs in the world, have helped many patients and continue to be the first-line pharmacological treatment for depression in most clinical settings. But the model is also, as has become increasingly clear over the past two decades, dramatically incomplete. The antidepressants derived from it typically take three to six weeks to produce clinical benefit, even though their biochemical effects on serotonin reuptake occur within hours. They are ineffective in approximately one-third of patients. And direct measurements of serotonin levels in depressed patients have not produced the consistent, dramatic deficiencies that the model predicts.

The emerging alternative, supported by a convergence of neuroimaging, electrophysiology, and pharmacological evidence, is a circuit disorder model: the proposal that depression is best understood as a disorder of specific neural circuits whose abnormal electrical activity patterns produce the characteristic symptoms of the condition. This is not a rejection of the neurochemical model but a more complete account of which circuits are affected, what their abnormal electrical behavior looks like, and how different treatments, pharmacological, psychological, and neuromodulatory, act to correct those abnormalities.

The Circuits of Depression

The most extensively studied circuit in the neuroscience of depression is the one that connects the prefrontal cortex, particularly the subgenual anterior cingulate cortex, with the amygdala, the hippocampus, and the subcortical reward structures of the basal ganglia. Each of these regions plays a specific role in the cluster of symptoms that constitute major depression, and the abnormal electrical activity in each region, and in the connections between them, can be measured and in some cases visualized with modern neuroimaging technology.

The subgenual anterior cingulate cortex, sometimes referred to by its Brodmann area designation as area 25, sits at the junction of the prefrontal cortex and the limbic system. It is heavily connected to the amygdala, the hippocampus, the hypothalamus, the brainstem, and the autonomic nervous system. In healthy individuals, area 25 shows moderate levels of activity that reflects its role in integrating emotional information and coordinating emotional responses. In people with major depression, area 25 is consistently hyperactive: it shows elevated metabolic activity and elevated electrical oscillations compared to healthy controls, and this hyperactivity appears to be a marker of the depressive state rather than a stable trait, since effective treatment of depression, whether by antidepressants, psychotherapy, electroconvulsive therapy, or deep brain stimulation, consistently reduces area 25 activity toward normal levels.

The amygdala, as described in Chapter Five, is the brain's alarm center. In depression, it shows hyperreactivity to negative stimuli and reduced habituation, the normal process by which repeated exposure to a stimulus leads to a diminishing response. A depressed person's amygdala responds more strongly to sad or threatening faces than a healthy person's does, and this elevated response persists longer and diminishes more slowly with repeated exposure. The result is a brain that is more readily alarmed by negative stimuli, more persistently distressed by them, and less able to return to a neutral emotional baseline after exposure to them.

The hippocampus, the seat of memory formation, shows a different kind of abnormality in depression. Structural neuroimaging studies, now replicated many times across many research groups, have consistently found that the hippocampus is smaller in people with major depression than in healthy controls, and that this volume reduction is proportional to the duration and severity of the depressive illness. The mechanism appears to involve the suppression of adult neurogenesis, the generation of new neurons in the hippocampus that occurs throughout adulthood and that is disrupted by the elevated glucocorticoid levels that accompany chronic stress and depression. Antidepressant drugs, and most other effective depression treatments, promote hippocampal neurogenesis, and some researchers believe this neurogenic effect is an important component of their therapeutic benefit.

The reward system, the dopaminergic circuitry of the basal ganglia and the nucleus accumbens, is the biological substrate of anhedonia, the inability to experience pleasure that is one of the most characteristic and most debilitating symptoms of major depression. In healthy individuals, anticipation of a reward activates the nucleus accumbens and produces a motivating rise in dopamine that drives approach behavior. In depressed individuals, this anticipatory dopamine response is blunted: the system is less responsive to reward, less able to generate the motivational signal that makes effort feel worthwhile, less capable of connecting the prospect of future pleasure to present action. This is why depressed people often know intellectually that they enjoy certain activities but cannot motivate themselves to engage in them: the circuit that connects knowledge to motivation to action has lost its electrical drive.

The Default Mode Network and Rumination

One of the most significant contributions of neuroimaging research to the understanding of depression in the past two decades has been the discovery and characterization of the default mode network: a collection of brain regions, including the medial prefrontal cortex, the posterior cingulate cortex, the precuneus, and the angular gyrus, that is more active during rest than during externally directed task performance. The default mode network was initially treated as something of an embarrassment by neuroscientists: the observation that a specific network of brain regions became more active when people were not doing anything in particular seemed to contradict the assumption that brain activity should increase in proportion to the cognitive demands being placed on it. Why should resting the brain increase the activity of a specific network?

The answer, which emerged gradually through a combination of neuroimaging and psychological research, is that the default mode network is not a resting network in any straightforward sense. It is the network of self-referential thought: the neural substrate of mind-wandering, daydreaming, remembering the past, imagining the future, and reflecting on the self and its social relationships. When you are not focused on an external task, your mind does not go blank. It wanders: it thinks about yourself, about your past, about other people, about possible futures. And this spontaneous, self-referential thinking is supported by the default mode network, which is doing its most characteristic work precisely when the brain is not engaged with the external world.

In people with major depression, the default mode network is chronically hyperactive and fails to suppress normally when external attention is demanded. The network that supports self-referential thinking, rumination, and the mental replay of negative experiences is running at too high a level and with too little regulatory restraint, flooding the depressed mind with an unrelenting stream of negative self-relevant thoughts and memories that constitutes one of the most recognizable and most distressing features of the depressive experience. The internal narrative of depression, the voice that repeats you are worthless, nothing will get better, you have always failed and you will always fail, has a neural substrate: it is the pathologically elevated activity of the default mode network, running its self-referential algorithms on a biased sample of experience, feeding its outputs into the hyperreactive amygdala, and maintaining the depressive state in a self-reinforcing loop that pharmacology alone is often insufficient to break.

Anxiety: The Alarm That Will Not Stop

Anxiety and depression are not the same condition, though they are deeply related and frequently co-occur. The majority of people with major depressive disorder also have significant anxiety, and the majority of those with anxiety disorders develop depression at some point. The neural overlap between the two conditions is substantial: both involve abnormalities in the amygdala, the prefrontal cortex, and the regulatory connections between them. But their specific electrical signatures are distinct, and understanding what distinguishes them is important for both scientific and clinical reasons.

Anxiety, at its neural core, is a disorder of threat detection and threat regulation. The amygdala's low-road threat detection system, described in Chapter Five, is calibrated to err on the side of over-detection: better to generate a fear response to a harmless stimulus than to fail to generate one for a genuinely dangerous one. In the environment in which the human amygdala evolved, the cost of false alarms was low and the cost of missed threats was potentially lethal. The result is a system with a built-in tendency toward over-activation, kept in check under normal circumstances by the regulatory influence of the prefrontal cortex, particularly the ventromedial prefrontal cortex, which sends inhibitory signals to the amygdala that dampen its responses to stimuli that experience has shown to be harmless.

Anxiety disorders, in this framework, represent a failure of prefrontal regulation to adequately modulate amygdala reactivity. The prefrontal regulatory signal is too weak, or the amygdala is too reactive, or both, and the result is an alarm system that fires too easily, too intensely, and for too long in response to stimuli that do not warrant the response. Generalized anxiety disorder involves a diffuse, chronic over-activation of this system across a wide range of stimuli. Specific phobias involve conditioned fear responses to particular classes of stimuli, maintained by the amygdala's extraordinary tenacity in holding onto fear associations and by the avoidance behaviors that prevent the extinction learning that would normally diminish them. Post-traumatic stress disorder involves particularly intense, persistent fear memories associated with a traumatic experience, encoded with the amplified emotional intensity that the amygdala's stress-hormone interactions with the hippocampus produce, and maintained by the same avoidance behaviors that prevent the neural processing that would allow the memories to lose their acute emotional charge.

The Hope on the Horizon

The man described at the beginning of this chapter is still searching. He has not found the treatment that gives him back the dimensionality he remembers from before the flatness came. He may find it. The science of depression and anxiety is advancing faster than at any previous point in the history of psychiatry, driven by the circuit disorder model, by new imaging technologies that can measure the brain's electrical activity with unprecedented spatial and temporal resolution, by the development of ketamine and esketamine as rapid-acting antidepressants that work through entirely different mechanisms than conventional antidepressants, by the growing evidence for psilocybin and MDMA in treatment-resistant cases, and above all by the development of neuromodulation technologies, deep brain stimulation, transcranial magnetic stimulation, closed-loop neurostimulation, that can target and correct the specific circuit abnormalities that drive the depressive state.

None of these advances has yet produced a treatment that works reliably for everyone, or one that is free from side effects, or one that is accessible to everyone who needs it regardless of their economic circumstances or the healthcare system they happen to inhabit. The science is ahead of the implementation. The biology is being understood faster than the clinical apparatus can translate that understanding into therapies. But the direction of travel is clear, and it is hopeful: toward treatments that are targeted to specific circuit abnormalities in specific individuals, delivered with sufficient precision to correct those abnormalities without disrupting the surrounding neural machinery, and adjusted in real time as the brain responds. The dark current is being illuminated. The question is whether the light can be brought to everyone who needs it.

CHAPTER TEN

The Forgetting Disease

Alzheimer's, Parkinson's, and the Electrical Collapse of the Aging Brain

SHE STILL RECOGNIZES the music. This is what her daughter notices, and notices with the particular quality of attention that belongs to people who are watching for any remaining thread of the person they knew in someone who is being slowly and inexorably unmade. The daughter's name is Priya, and her mother has had Alzheimer's disease for six years. In those six years, her mother has lost, in the order in which Alzheimer's characteristically takes them, her most recent memories first, then her ability to follow a complex conversation, then her understanding of dates and times and why she is in this room rather than another, then the faces of people she has known for less than forty years, then a succession of words that left her speech full of gaps she tried to paper over with gestures and approximations, then the ability to prepare food safely, then the name of her daughter.

But she still recognizes the music. Priya plays a recording of the film songs her mother loved in the 1970s and 1980s, the songs she sang while cooking, while driving, while doing the particular tasks of an ordinary life that she can no longer perform, and her mother's face changes. Not dramatically. Not a restoration. But a loosening of the blankness that has settled over the features, a softening, a tentative emergence of something that is recognizably her. She does not know where she is or who Priya is. But she knows this music, and in knowing it she is, for these moments, more fully herself than at almost any other time of the day.

This is not a miracle. It is neuroscience. The musical memories her mother retains are stored in a different part of the brain, with a different electrical and anatomical organization, than the declarative memories that Alzheimer's disease has systematically destroyed. The disease has followed its characteristic anatomical progression, beginning in the hippocampus and the entorhinal cortex and moving outward through the association areas of the temporal and parietal lobes, destroying the circuitry of episodic and semantic memory while leaving intact for much longer the emotional and procedural systems, including the musical memory systems of the auditory cortex and the limbic structures associated with emotional memory, that the disease reaches later and affects differently. Understanding why is not merely scientifically interesting. It is the foundation for understanding how to delay, treat, and eventually prevent one of the most devastating conditions that human beings endure.

What Alzheimer's Disease Is

Alzheimer's disease is a progressive neurodegenerative condition characterized by the loss of neurons and synapses in specific brain regions, driven by the accumulation of two abnormal protein deposits: amyloid-beta plaques in the spaces between neurons and tau tangles inside neurons. It is the most common cause of dementia, accounting for sixty to seventy percent of all cases, and it affects an estimated fifty-five million people worldwide, a number projected to nearly triple by 2050 as global populations age. It is the sixth leading cause of death in the United States, and unlike most major diseases, its mortality rate has been increasing rather than declining over the past two decades, reflecting both its increasing prevalence and the continued inadequacy of available treatments to significantly alter its course.

The disease was first described in 1906 by the German psychiatrist and neuropathologist Alois Alzheimer, who presented the case of a fifty-one-year-old woman named Auguste Deter at a medical conference in Tubingen. Deter had developed a progressive dementia characterized by memory loss, disorientation, hallucinations, and behavioral changes in her late forties, an unusually early age for dementia onset, which is why Alzheimer originally described it as a presenile dementia, distinguishing it from the more common senile dementia of old age. When Deter died in 1906, Alzheimer performed a detailed neuropathological analysis of her brain and identified the two structural abnormalities that still bear his name as the diagnostic hallmarks of the disease: the amyloid plaques and the neurofibrillary tangles.

For the first seven decades after Alzheimer's original description, the condition that bore his name was considered a rare disease affecting people in middle age. It was the work of Robert Katzman and Robert Terry in the 1970s that established the continuity between early-onset Alzheimer's disease and the far more common late-onset dementia of old age, showing that both conditions shared the same neuropathological features and were, at the biological level, the same disease. This reclassification, which transformed Alzheimer's disease from a medical curiosity to a public health crisis almost overnight, is one of the most consequential contributions to the epidemiology of aging in the twentieth century.

The Amyloid Hypothesis and Its Challenges

The dominant scientific theory of Alzheimer's disease for the past three decades, the theory that has guided the vast majority of clinical drug development and the allocation of research funding, is the amyloid cascade hypothesis. First proposed in its modern form by John Hardy and David Allsop in 1991 and refined by Hardy and Gerald Higgins in 1992, the hypothesis proposes that the accumulation of amyloid-beta in the brain is the primary initiating event in Alzheimer's pathology, and that all subsequent developments, the tau pathology, the neuroinflammation, the synaptic dysfunction, the neuronal death, and the clinical symptoms, are downstream consequences of amyloid accumulation.

The amyloid cascade hypothesis is supported by compelling genetic evidence. All of the known rare genetic mutations that cause early-onset familial Alzheimer's disease, occurring in people in their thirties, forties, and fifties, are in genes that directly affect amyloid-beta production or metabolism. People with Down syndrome, who carry an extra copy of chromosome 21, which contains the gene for amyloid precursor protein, develop Alzheimer's pathology in their thirties and clinical dementia in their fifties at dramatically elevated rates. And the genetic variant most strongly associated with late-onset sporadic Alzheimer's disease, the APOE4 allele of the apolipoprotein E gene, affects the brain's ability to clear amyloid-beta, leading to its accumulation over time.

But the hypothesis has also faced a series of serious challenges that have complicated the picture considerably. The most damaging have come from clinical trials: numerous drugs designed to reduce amyloid levels in the brain have been developed over the past two decades and tested in large clinical trials, and the vast majority have failed to produce meaningful clinical benefit in patients with established Alzheimer's dementia, even when they successfully reduced amyloid burden as measured by brain imaging. The trials of drugs like solanezumab, gantenerumab, crenezumab, and many others produced anti-amyloid effects without clinical benefits, raising the uncomfortable possibility that by the time patients have dementia, amyloid reduction may simply come too late to reverse the downstream damage.

More recent trials of more potent anti-amyloid antibodies, particularly lecanemab, approved by the FDA in 2023, and donanemab, have produced statistically significant but clinically modest slowing of decline in patients with early-stage disease, providing the first genuine evidence that anti-amyloid treatment can affect the clinical course of the disease if begun early enough. These results have reinvigorated the amyloid hypothesis and established the principle that earlier treatment, before significant neuronal loss has occurred, may be essential for any anti-amyloid approach to succeed.

The Electrical Collapse

Whatever the initiating cause of Alzheimer's disease, and whatever the relative contributions of amyloid, tau, neuroinflammation, mitochondrial dysfunction, and the many other factors that have been implicated, the functional consequence that produces the clinical syndrome of dementia is an electrical collapse: the progressive loss of the synaptic connections and neural circuits that enable the cognitive functions of memory, language, reasoning, and self-awareness. Understanding this electrical collapse, and finding ways to detect, measure, and potentially reverse it, is one of the most active areas of Alzheimer's research.

The earliest electrical changes in Alzheimer's disease, detectable years or even decades before any clinical symptoms appear, involve the hippocampus and the entorhinal cortex, the gateway through which information enters and exits the hippocampal memory system. Neurons in the entorhinal cortex, called grid cells and place cells, fire in patterns that create a cognitive map of the environment, and the integrity of these firing patterns depends on the precise synaptic connectivity of the hippocampal-entorhinal circuit. In people who carry the genetic risk for Alzheimer's disease and in those with early amyloid accumulation but no clinical symptoms, subtle changes in the activity patterns of the hippocampus can be detected using high-field fMRI: the hippocampus is paradoxically hyperactive in these early stages, as if working harder to compensate for the early loss of synaptic efficiency.

As the disease progresses, this early hyperactivity gives way to progressive hypoactivity and eventual silence as neurons are lost and their connections are destroyed. The characteristic EEG changes of established Alzheimer's disease, a slowing of the dominant cortical rhythms from the normal alpha frequency range into the theta and even delta range, reflect this progressive loss of the high-frequency synchronization that depends on the rapid, precise synaptic connections that the disease has destroyed. The pattern of EEG slowing correlates with the severity of cognitive impairment and follows the anatomical progression of the disease, beginning in the temporal regions and spreading progressively to the parietal and frontal lobes as the pathology advances.

There is one particularly fascinating aspect of this electrical collapse that brings us back to Priya's mother and her preserved musical memory. The regions of the brain responsible for musical memory, including the auditory cortex, the motor areas involved in rhythm and movement, and the limbic structures associated with the emotional response to music, are among the last to be affected by Alzheimer's pathology. The anatomical progression of the disease, which follows the connectivity of the entorhinal-hippocampal system outward through the association cortices, reaches these regions only in the later stages of the disease. Musical memories, which are encoded differently from episodic and semantic memories and which are supported by this more distributed, emotion-linked network, therefore survive longer than the memories that the disease attacks first.

Parkinson's Disease: The Motor Circuit

Parkinson's disease is in many respects the mirror image of Alzheimer's. Where Alzheimer's disease attacks primarily the cognitive and memory systems of the cortex and hippocampus, Parkinson's disease attacks primarily the motor control systems of the basal ganglia, producing the characteristic triad of tremor, rigidity, and bradykinesia, the slowness and poverty of movement that gives the disease its name from the description of the shaking palsy published by the English physician James Parkinson in 1817.

At the heart of Parkinson's disease is the progressive degeneration of a small population of neurons in the substantia nigra, a darkly pigmented region in the midbrain that produces dopamine and sends it via the nigrostriatal pathway to the striatum, the major input structure of the basal ganglia. The basal ganglia are, among other things, a motor gating system: they regulate which movements are initiated and which are suppressed, selecting the appropriate motor programs for the current situation and inhibiting competing programs that would interfere with smooth, coordinated movement. The dopamine signal from the substantia nigra is essential for this selection process: without it, the basal ganglia's inhibitory output becomes excessive and indiscriminate, suppressing movements that should be facilitated and failing to adequately inhibit movements that should be suppressed. The result is the motor poverty of Parkinson's disease: difficulty initiating movement, slowness and reduction in the amplitude of movement, and the involuntary tremor that paradoxically accompanies the poverty of intentional movement.

The electrical consequences of dopamine loss in the basal ganglia have been studied in exquisite detail using the recording equipment implanted in patients during deep brain stimulation surgery, the most effective treatment currently available for the motor symptoms of Parkinson's disease. What these recordings reveal is that the dopamine-depleted basal ganglia develop an abnormal pattern of synchronized oscillation in the beta frequency range, approximately thirteen to thirty hertz, that appears to suppress the motor system in a way that produces the characteristic features of Parkinsonism. In healthy individuals, beta synchronization in the motor circuits decreases when movement is about to begin, in a phenomenon called beta event-related desynchronization that reflects the motor system's preparation for action. In Parkinson's disease, this desynchronization is impaired: the basal ganglia remain locked in excessive beta synchrony that continues to suppress movement even when the patient is trying to move.

Deep brain stimulation of the subthalamic nucleus, one of the key nodes in the basal ganglia circuit, is extraordinarily effective at suppressing this pathological beta synchrony and restoring more normal patterns of motor circuit activity. When the stimulator is turned on, the beta synchrony drops, the motor suppression is relieved, and movement becomes more fluid and more easily initiated. When the stimulator is turned off, the beta synchrony returns and the Parkinson's symptoms re-emerge. The treatment is not a cure: the underlying neurodegeneration continues regardless of the stimulator's operation, and the disease eventually progresses to involve additional circuits that deep brain stimulation does not address. But for the motor symptoms that are its primary target, it is one of the most effective and most dramatic treatments in all of neurology, providing a level of motor function that would otherwise be impossible and giving patients back a quality of life that the disease would otherwise have taken from them.

The Path Forward

Priya's mother does not know the name of the disease that is remaking her. She does not know that her brain is being unmade, neuron by neuron, synapse by synapse, in a process that science understands incompletely and medicine can only slightly and temporarily slow. She knows the music. She knows, in some register that goes deeper than episodic memory, that this sound is meaningful and that it is hers. She knows, perhaps, something of the warmth of the person sitting beside her, though she cannot name her.

For Priya, and for the tens of millions of people around the world who are living through this with someone they love, the science of Alzheimer's disease is not an academic matter. It is the most urgent possible question: can we stop this? The honest answer, in 2026, is: not yet, not completely, but we are closer than we have ever been. The recent clinical trial successes with anti-amyloid antibodies, modest as they are, have established the principle that earlier intervention can slow the disease's course. The development of blood-based biomarkers that can detect amyloid and tau pathology years or decades before symptoms appear is opening the door to treatment at a stage when the electrical architecture of the brain is still sufficiently intact that preserving it is possible. The research into tau therapies, neuroinflammation, the glymphatic system, and the role of sleep in amyloid clearance is generating new targets and new approaches that could, in the next decade, produce treatments of substantially greater efficacy than any currently available.

The brain that is being unmade is not all of the person. Priya knows this. The electrical catastrophe does not reach everything at once. It proceeds along its anatomical pathways, and in the regions, it has not yet reached, the person her mother is continues to exist, diminished but present, responding to music in the brain's most ancient emotional registers, knowing something of the warmth of a daughter's presence even without the name to attach to the face. The science that is trying to stop the catastrophe is, in the meantime, also the science that is helping families and caregivers understand what they are seeing and why, and to find in that understanding something that is not hope exactly, but not hopelessness either. Something closer to clarity: the clear-eyed recognition of what is happening and what remains, and the resolution to be present for both.

PART FOUR

REWIRING THE BRAIN

Electrical and Chemical Tools for Healing and Enhancing the Neural Circuits

Part Introduction

The Age of Intervention

The chapters of Part Three described what happens when the brain's electrical machinery fails: the synchronization catastrophes of epilepsy, the circuit dysregulation of depression and anxiety, the progressive electrical collapse of Alzheimer's and Parkinson's disease. Part Four describes what happens when medicine and neuroscience attempt to fix it. The story of how we learned to intervene in the brain's electrical life, to send targeted electricity in through implanted electrodes, to read the brain's signals and decode the thoughts they carry, and to chemically rewire its habitual patterns through compounds that science is only beginning to understand, is one of the most extraordinary stories in the history of medicine.

It is also a story with a particular structure. Each of the interventions described in the next three chapters began in a place that was not quite where it ended up. Deep brain stimulation began as an accident, a surgical observation made during a procedure for a different purpose, that gradually became the most effective treatment for Parkinson's disease and a promising therapy for half a dozen other conditions. The decoding of neural signals began as a basic science effort to understand how motor commands are encoded in neural activity, and became, within two decades, the technology that allowed paralyzed people to type with their thoughts and speak with their minds. And the psychedelic medicines began as tools of psychiatry, were banned for thirty years as tools of counterculture, and have returned as perhaps the most promising class of psychiatric treatment in half a century, their mechanism of action only now being understood in terms of what they do to the brain's electrical organization.

Part Four is, in its deepest structure, a meditation on what it means to intervene in the brain. The brain is not like other organs. When you repair a damaged heart valve, you restore function without altering the person whose heart it is. When you treat a brain with electricity or chemistry or the dissolution of its habitual patterns, you are doing something more intimate, more complex, and more fraught. You are changing the organ that makes the person who they are. The successes of neural intervention are extraordinary. The questions they raise about identity, autonomy, and the boundaries of the self are equally extraordinary. Both deserve the attention of anyone who wants to understand where the neuroscience of the electric brain is taking us.

CHAPTER ELEVEN

Sending Electricity In

Deep Brain Stimulation, TMS, and the Electrical Healing of Broken Circuits

ON A THURSDAY AFTERNOON in the spring of 2003, a woman named Helen sat in a neurosurgeon's office in Toronto and listened to a description of what was about to happen to her brain. The description was careful and honest and included, among other things, the information that the procedure she was considering had never been performed for her specific condition in the way her surgeon was proposing, that the evidence base for it was a series of case reports rather than a randomized controlled trial, and that the most likely outcome was that it would not work. Helen had been profoundly depressed for most of her adult life. She had tried eleven antidepressants, two courses of electroconvulsive therapy, and years of psychotherapy. Nothing had produced more than a temporary and partial improvement. She was, in the clinical terminology that does not quite capture the reality of what she was describing when she tried to explain it to other people, treatment-resistant. She was also, she would say later, simply tired of being the way she was. She said yes.

The surgeon was Andres Lozano, one of the most skilled and innovative functional neurosurgeons in the world, and his colleague in the psychiatric side of this collaboration was Helen Mayberg, a neurologist and neuroscientist at the University of Toronto whose neuroimaging research had identified the subgenual anterior cingulate cortex, area 25, as a critical node in the circuit of depression. As described in Chapter Nine, area 25 is chronically hyperactive in major depression and its activity normalizes with successful treatment by any method. Mayberg had developed the hypothesis that directly targeting area 25 with deep brain stimulation might correct this hyperactivity in patients who had failed to respond to every other available treatment.

The surgical procedure itself took several hours. Under local anaesthesia, with Helen awake throughout, Lozano drilled two small holes in her skull and passed thin electrode leads deep into the brain, positioning their tips in the white matter adjacent to area 25, guided by preoperative MRI images and by the real-time feedback of electrical recordings made from the electrode tips during placement. The leads were connected to an implantable pulse generator, a device roughly the size of a matchbox that would be placed beneath the skin of the chest wall and that would deliver continuous electrical stimulation to the target site, day and night, for as long as it was needed.

When the stimulator was turned on for the first time, with Helen awake in the operating room, something happened that neither Lozano nor Mayberg had quite expected, despite everything they knew about the neuroscience. Helen, who had been sitting quietly in the characteristic stillness of severe depression, began to describe changes almost immediately. The feeling of pressure around her head that she had not previously mentioned because it had been present for so long she had stopped registering it as a symptom, was gone. The room, which had seemed to her the flat, dimensionless gray of the depressive world, seemed suddenly to have recovered color and depth. She said, in words that Mayberg would quote in lectures for years afterward, it is like the heaviness is lifting. Within days, she was substantially better. Within weeks, her depression was in remission for the first time in years.

A Discovery Made by Accident

Deep brain stimulation did not begin with depression. It began, as many of the most important discoveries in medicine do, with an observation made during a procedure undertaken for an entirely different purpose, an observation that was strange enough and significant enough that the surgeon who made it could not ignore it and spent the following years trying to understand it.

The surgeon was Alim Louis Benabid, a French neurosurgeon at the University of Grenoble, and the year was 1987. Benabid was performing a thalamotomy, a surgical procedure in which a lesion is created in the thalamus to treat the tremor of Parkinson's disease, a procedure that was standard practice at the time and that was effective but irreversible. Before creating the lesion, he used electrical stimulation to confirm the placement of his electrode, a routine step in the procedure to ensure that the electrode was in the right location. He stimulated at the low frequency of 50 hertz, a standard test frequency, and noted the expected response. Then, on an instinct he later described as clinical intuition, he increased the frequency to 100 hertz. The patient's tremor, which had been present throughout the procedure, stopped.

Benabid had observed something that his training had not prepared him to expect: that high-frequency electrical stimulation of the thalamus suppressed tremor as effectively as a surgical lesion, but reversibly, and without destroying tissue. When the stimulation was turned off, the tremor returned. When it was turned on, the tremor stopped. The effect was immediate, reliable, and completely reversible. He had, in effect, discovered a way to create a functional lesion without creating a structural one: to achieve the therapeutic effect of surgical tissue destruction through the alternative mechanism of electrical disruption of neural activity, a disruption that ceased the moment the electrical current was turned off.

Benabid spent the next several years refining the technique, identifying the optimal target for Parkinson's tremor as not the thalamus but the subthalamic nucleus, a small but critically important node in the basal ganglia circuit, and collaborating with the device manufacturer Medtronic to develop implantable stimulators suitable for long-term clinical use. The first large clinical trial of deep brain stimulation for Parkinson's disease, published in the New England Journal of Medicine in 2001, was transformative: patients who had been severely disabled by tremor, rigidity, and bradykinesia showed dramatic improvements in motor function with the stimulator on, and returned to their disabled state when the stimulator was turned off. The FDA approved deep brain stimulation for Parkinson's disease in 2002, and it has since been performed in more than 200,000 patients worldwide, making it one of the most widely used and most successful neurosurgical procedures in history.

How Deep Brain Stimulation Works

The mechanism by which deep brain stimulation produces its therapeutic effects has been, somewhat embarrassingly for a field that prides itself on evidence-based practice, debated and incompletely understood for most of the three decades since Benabid's initial observation. The original hypothesis was that high-frequency stimulation inhibited the target nucleus, producing an effect equivalent to a lesion. This hypothesis had the virtue of explaining why DBS worked: if destroying the subthalamic nucleus helped Parkinson's, then inhibiting it should help too. But it had the defect of being, in important respects, wrong.

More sophisticated electrophysiological studies, using microelectrode recordings made in patients during DBS surgery and in animal models, have revealed that the actual mechanism is considerably more complex. High-frequency stimulation does not simply inhibit the target nucleus. It overrides the normal patterned activity of the target neurons, replacing the pathological synchronized oscillations with a high-frequency, more randomized pattern of firing that disrupts the abnormal circuit dynamics without simply silencing the circuit. In the case of Parkinson's disease, this disruption of the pathological beta-frequency synchrony in the basal ganglia, described in Chapter Ten, releases the motor system from its pathological suppression and allows more normal movement to proceed.

This more nuanced understanding of DBS mechanisms has been enormously useful for extending the technique to new applications. If DBS works by disrupting pathological neural oscillations rather than simply inhibiting neural activity, then any condition that involves identifiable pathological oscillations in specific circuits is potentially a candidate for DBS therapy. Depression, with its hyperactive area 25 and its pathologically elevated default mode network activity. Obsessive-compulsive disorder, with its hyperactive cortico-striato-thalamic loops. Epilepsy, with its paroxysmal hypersynchrony. Chronic pain, with its abnormal sensory processing circuits. The list of conditions for which DBS has been investigated, with varying degrees of evidence, has grown considerably in the two decades since its initial approval, and the list of conditions for which it has achieved sufficient evidence to justify clinical use continues to expand.

Stimulation Without Surgery

Deep brain stimulation requires surgery. It requires drilling holes in the skull, passing electrodes through living brain tissue, and implanting a device that will remain in the body for years or decades. For many patients with severe, treatment-resistant conditions, this level of intervention is justified by the magnitude of the potential benefit. But for the much larger population of patients with moderate depression, anxiety, chronic pain, or other conditions that might benefit from some form of neuromodulation, a non-surgical alternative is highly desirable.

Transcranial magnetic stimulation, TMS, provides exactly this. The technique uses a coil of wire placed against the skull through which a powerful, rapidly changing magnetic field is passed. This changing magnetic field induces an electrical current in the brain tissue beneath the coil, sufficient to depolarize neurons and cause them to fire, without any physical contact between the stimulating equipment and the brain. The depth of stimulation is limited to approximately two to three centimetres below the skull surface, which means that TMS can effectively target the cortex but not the deep subcortical structures accessible to DBS. But for depression, whose cortical components include the hyperactive default mode network and the dysregulated prefrontal cortex, this limitation is acceptable.

The FDA approved repetitive TMS for the treatment of major depression in 2008, making it the first non-pharmacological, non-surgical depression treatment to receive regulatory approval since electroconvulsive therapy. The typical treatment course involves daily sessions of TMS delivered to the left dorsolateral prefrontal cortex for four to six weeks. The mechanism is still not fully understood, but the evidence is clear: approximately fifty to sixty percent of patients who have not responded to at least one antidepressant medication show a significant clinical response to TMS, and approximately thirty to thirty-five percent achieve full remission. These response rates compare favorably with those of antidepressant medications, and the side effect profile is substantially more benign: the most common side effects are headache and scalp discomfort at the stimulation site, without the weight gain, sexual dysfunction, emotional blunting, and discontinuation syndrome that accompany many antidepressant medications.

More recently, the development of accelerated TMS protocols, in which multiple treatment sessions are delivered each day over a compressed time period, has produced response rates that appear substantially higher than the standard protocol, with some studies reporting remission rates above sixty percent in treatment-resistant patients. And the development of theta burst stimulation, a TMS protocol that delivers stimulation in a pattern designed to mimic the brain's endogenous theta rhythms and that can be administered in three minutes rather than the forty minutes of standard TMS, has made high-volume treatment schedules logistically feasible in ways they previously were not.

The Question of Identity

Helen, the woman described at the beginning of this chapter, has now lived with her DBS device for more than two decades. By any clinical measure, her outcome has been extraordinary: her depression has been in sustained remission, interrupted only by the periods when her device's battery has needed replacement and there has been a gap between the battery's exhaustion and the surgical replacement. During those gaps, she has felt the depression return, with a speed and a completeness that makes clear, if there were any doubt, that it is the electrical intervention that is keeping her well.

But Helen has also, over the years, found herself thinking about questions that the neuroscience of her situation raises but does not answer. When the device is keeping her well, when she is cheerful and energetic and engaged with life in ways that she was not before the surgery, is she experiencing her authentic self? Or is she experiencing a self that the device is creating, a state that is better than her natural state but is therefore, in some sense, not entirely her own? If the device were turned off tomorrow and she returned to the depression that is, in some neurobiological sense, her brain's default state without intervention, which of these states represents the real Helen?

There are no easy answers to these questions, and it would be a mistake to pretend otherwise. What can be said is that the questions reveal something important about the brain: that what we call the self is not a fixed, given thing but a state of the brain, a particular pattern of neural activity that can be more or less well-functioning, more or less consistent with the person's values and capacities and sense of their own identity. When a brain disorder disrupts that state, as depression disrupts Helen's, it does not reveal a truer self beneath the illness. It creates a different, worse-functioning state that the person does not recognize as themselves. And when an intervention restores the better-functioning state, it is not creating an artificial self. It is restoring the one that was there before the disorder took it. That, at least, is the most defensible position that the science supports. The deeper philosophical question, whether there is any fact of the matter about which brain state represents the authentic self, remains genuinely open. Helen lives with the question. Most of the time, with the device running, she is too busy living to worry about it very much. This seems, to her and to most who know her, like the right answer.

CHAPTER TWELVE

Reading the Electric Mind

Brain-Computer Interfaces and the Decoding of Thought

IN MAY OF 2023, A PAPER appeared in the journal Nature Neuroscience that described something that, a decade earlier, would have seemed the premise of a science fiction novel rather than a report of clinical reality. A team of researchers at the University of Texas at Austin, led by the cognitive neuroscientist Alexander Huth, had developed a system that could reconstruct the continuous semantic content of what a person was hearing, or imagining, from the patterns of brain activity recorded by functional MRI. The system was not perfect. It did not reproduce exact words or sentences. But it captured the meaning of what a person was hearing with an accuracy that far exceeded chance, reconstructing the gist of a narrative from brain activity in a way that an independent reader could recognize as reflecting the content of the original stimulus. For the first time in the history of neuroscience, a practical system existed for decoding the semantic content of a person's thoughts from their brain activity alone.

The headline that accompanied almost every news report on the paper was, predictably, Scientists Read Minds. The reality was more precise and more interesting than that headline suggests, and also more consequential, because the precise reality is one that has implications for privacy, for autonomy, for the law, and for what it means to have a private inner life in a world where the technology of neural decoding is advancing at its current pace. Understanding what was actually achieved, what it means, and what it foreshadows, is the task of this chapter.

The Road to BrainGate

The systematic effort to decode the brain's motor signals and use them to control external devices began not with a grand vision of restoring communication to paralyzed patients but with a basic science question about how the motor cortex represents movement. In the 1980s and 1990s, researchers including Apostolos Georgopoulos at the Johns Hopkins University School of Medicine were recording the electrical activity of individual neurons in the motor cortex of macaque monkeys while the animals performed reaching movements in various directions. They found that each motor neuron had a preferred direction: it fired most strongly when the arm moved in a particular direction, and fired less strongly as the direction of movement deviated from that preferred direction. By recording from many neurons simultaneously and summing their contributions weighted by their preferred directions, Georgopoulos could reconstruct the direction of the monkey's intended arm movement from the neural population activity alone, before the movement began.

This was the proof of concept that made brain-computer interfaces conceivable as a practical technology: the demonstration that the motor cortex encodes intended movements in a population code that can, in principle, be read out from recordings of neural activity. The next step was to show that this readout could be used to control an external device in real time, closing the loop between the brain's intention and the device's action.

That step was taken by the BrainGate consortium, a collaboration between Brown University, Massachusetts General Hospital, and several other research institutions, in a landmark experiment published in Nature in 2004. Their subject was Matthew Nagle, a twenty-four-year-old man from Weymouth, Massachusetts, who had been paralyzed from the neck down by a spinal cord injury sustained in a knife attack in 2001. Nagle had volunteered to have a small array of 96 microelectrodes implanted in the primary motor cortex of his brain, recording the activity of approximately 100 neurons simultaneously. After calibrating the system by having Nagle imagine making movements and recording the corresponding neural patterns, the researchers connected the neural recording system to a computer cursor. Nagle, using only his thoughts, was able to move the cursor across the screen, navigate a web page, adjust the volume of a television, play a primitive video game, and draw simple shapes. He could not type, precisely. He could not restore the full dexterity of his lost hands. But he could control a computer with his mind, in real time, for the first time in history.

From Cursor to Conversation

The two decades of BrainGate research that followed Nagle's landmark experiment produced a succession of capabilities that each seemed, at the time of their demonstration, nearly miraculous and that now, in retrospect, trace a coherent trajectory toward the restoration of full communicative capacity in people with profound paralysis. Subjects with increasingly severe motor impairments, including people with amyotrophic lateral sclerosis who had lost the ability to move any voluntary muscle in their bodies, were implanted with increasingly sophisticated electrode arrays and demonstrated increasingly powerful control over external devices.

The rate of progress accelerated dramatically in the late 2010s and early 2020s, driven by advances in machine learning algorithms for decoding neural signals, by improvements in electrode technology and wireless data transmission, and by the extraordinary motivation of the researchers and the remarkable courage and persistence of the patients who participated in clinical trials that required having recording equipment implanted in their brains with no guarantee of benefit and no historical precedent for what they were attempting.

In 2021, a paper in the New England Journal of Medicine described a BrainGate participant who had been paralyzed by ALS for years and who, using a handwriting BCI system that decoded the neural signals of imagined handwriting movements, could type at ninety characters per minute, faster than the previous record for any BCI system, with an accuracy of ninety-four percent after autocorrect. The speed and accuracy were sufficient for genuine, fluent communication in real time, not the laborious letter-by-letter spelling of earlier systems but something approaching the pace of natural written communication. He could compose and send emails. He could answer questions in interviews. He could write.

The milestone that generated perhaps the greatest public response came in 2023, when a team at the University of California at San Francisco, led by the neurosurgeon Edward Chang, reported in the journal Nature the restoration of speech to a woman named Ann, who had been unable to speak for eighteen years following a brainstem stroke. Electrodes implanted on the surface of Ann's speech motor cortex recorded the neural signals associated with her attempts to speak. A machine learning system trained on these signals, using Ann's own pre-injury voice recordings as a reference, decoded her intended speech at a rate of seventy-eight words per minute and synthesized it as an audio signal through a digital avatar that animated in synchrony with the decoded speech. The accuracy was approximately seventy-eight percent, with some phonemes and words more reliably decoded than others. When the research team played a recording of the demonstration to other scientists, the reaction was uniformly described as one of emotional intensity that is unusual in scientific contexts. Hearing a person speak who has not been able to speak for eighteen years, through technology that reads the speech directly from the electrical activity of her brain, is not an experience that anyone present described as merely scientific.

Decoding Perception and Language

The BrainGate work and its successors address what might be called the output side of the brain-computer interface problem: decoding motor intentions from the motor cortex to control external devices. But the University of Texas work described at the opening of this chapter addresses something different and, in some respects, more philosophically consequential: the input side, the decoding of perception and thought from the brain regions that process sensory experience and language rather than from the regions that generate movement.

Alexander Huth's research uses functional MRI to record brain activity while subjects listen to extended narratives, hours of podcast content, and then trains machine learning models to find the mapping between the patterns of brain activity and the semantic content of what is being heard. The models are then tested on brain activity recorded while subjects hear new, previously unseen narratives, and they are asked to reconstruct the semantic content of those narratives from the brain activity patterns alone. The reconstructions are not transcriptions. They are not verbatim reproductions of the spoken words. But they capture the meaning with a fidelity that demonstrates something that was not previously demonstrable: that the semantic content of language comprehension is represented in the brain in patterns that are consistent across individuals and that can be decoded from the outside.

The earlier work of Jack Gallant, also at Berkeley, established related principles for visual imagery: that the patterns of brain activity evoked by viewing natural images contain enough information about the content of those images to allow reconstruction of what a person was looking at, from the brain activity alone, without any direct access to the visual stimulus. The reconstruction quality improves as the machine learning models become more sophisticated and as the brain activity measurements become more precise. It is not yet possible to reconstruct visual experience with high fidelity from non-invasive brain measurements. But the trajectory of improvement over the past decade makes it clear that the fundamental limitation is technological rather than biological: the information about visual experience is present in the brain activity, and more precise measurements and more powerful decoding algorithms will continue to extract more of it.

The Privacy Crisis That Is Coming

The technology of neural decoding is advancing faster than the legal and ethical frameworks needed to govern it. This is not unusual in the history of transformative technology. The legal frameworks governing digital privacy were decades behind the technology that created the privacy problems they were meant to address. The legal frameworks governing genetic information were years behind the technology that made genetic testing cheap and ubiquitous. The legal frameworks governing neural data are, currently, almost entirely absent, and the technology of neural decoding is approaching the point where this absence will matter urgently.

Consider what the combination of currently existing technologies implies for the future of neural privacy. Implanted brain-computer interfaces can already record neural activity with millisecond temporal resolution and electrode-level spatial resolution from hundreds or thousands of recording sites simultaneously. Non-invasive wearable EEG devices are becoming smaller, cheaper, and more capable. Consumer-grade EEG headsets are already available for a few hundred dollars and record enough neural information to decode, with current algorithms, significant information about a user's emotional state, cognitive load, attention, and in some cases specific thoughts and intentions. The machine learning algorithms for decoding this neural information are improving rapidly, driven by the enormous investments in artificial intelligence research that are transforming every field that involves the analysis of complex high-dimensional data.

The combination of accessible neural recording technology and powerful decoding algorithms creates a situation that has no precedent in human history: the possibility that a person's thoughts, intentions, emotional states, and perceptual experiences could be read from the outside, without their explicit cooperation, from neural signals that they cannot voluntarily suppress or conceal the way they can suppress or conceal words or facial expressions. This is not a distant future possibility. It is a near-term practical concern, and it demands a legal and ethical response that has barely begun to be developed.

Some jurisdictions are beginning to respond. Chile amended its constitution in 2021 to include explicit protections for mental privacy, mental integrity, and what the framers of the amendment called neurorights, the rights of individuals to control access to their own neural data. The philosopher and neuroscientist Rafael Yuste at Columbia University, who was instrumental in developing the neurorights framework and in advocating for the Chilean constitutional amendment, argues that without explicit legal protection of neural data, the electric self will become the most exploited resource in human history: more intimate than medical records, more revealing than any other source of personal information, and without protection, available to whoever can access the technology to read it.

The questions Yuste and others are raising are not science fiction. They are policy questions being actively debated in legislatures and international bodies right now, in parallel with the scientific progress that makes them urgent. Who owns the data generated by a brain-computer interface? The patient who provided their neural activity? The company that made the device? The healthcare system that provided the implantation? Can neural data be subpoenaed in criminal or civil proceedings? Can it be sold to advertisers? Can insurance companies access it? Can employers? Can governments? These questions have not been answered, and the technology is not waiting for the answers to arrive before it advances.

The Hope Side

It would be a mistake to end this chapter on the alarm of the privacy crisis without emphasizing, with equal clarity, the extraordinary hope that the same technology represents for the people whose lives it is already beginning to transform. Ann, who speaks through her avatar. The man who writes emails with his imagined handwriting. The thousands of Parkinson's patients whose tremors are controlled by stimulators that read their neural state and adjust their output accordingly. The paralyzed patients who are, in research settings right now, operating robotic arms with sufficient dexterity to grasp a cup of coffee and bring it to their lips.

For these people, the technology of reading the electric mind is not a privacy threat. It is a liberation. It is the restoration of a form of agency, of a connection to the world and to other people, that the disease or the injury took from them. The same technology that could, in the wrong hands and without adequate regulation, be used to surveil and exploit and control, is also, in the right context and with the right safeguards, one of the most compassionate and consequential medical achievements in the history of human beings trying to help each other. Both of these things are true simultaneously. The technology is not good or bad. The choices about how to develop, govern, and deploy it are good or bad. And those choices, which are being made right now, will determine whether the age of reading the electric mind is remembered as a liberation or a violation. Or, most likely, as both at once, in the complicated way that transformative technologies almost always are.

CHAPTER THIRTEEN

The Psychedelic Rewire

How Ancient Molecules Dissolve the Brain's Rigid Electrical Habits and Create New Ones

IN THE SPRING OF 2016, a sixty-eight-year-old woman named Dinah, who had smoked cigarettes for forty-five years and failed multiple previous quit attempts with nicotine replacement therapy, varenicline, and hypnosis, took a capsule of psilocybin in a carefully arranged room at Johns Hopkins University School of Medicine, lay down on a comfortable couch with eyeshades covering her eyes and carefully selected music playing through headphones, and had an experience that she later described as the most meaningful of her life. She reported feeling the interconnection of all living things with a completeness and an emotional certainty that she had never previously experienced. She reported a conversation with a presence she could not name but that seemed, in the logic of the experience, to represent something more essential than herself. She reported a complete dissolution of the boundary between herself and the world around her that lasted for several hours and that was accompanied by a quality of significance and peace that was unlike anything in her forty-five years of ordinary waking experience. When she woke from the experience, she had no craving for cigarettes. At the twelve-month follow-up, she had not smoked since the session.

Dinah is one of fifteen participants in the first clinical study of psilocybin-assisted therapy for smoking cessation, published in

Psychopharmacology in 2014, in which eighty percent of participants were abstinent from smoking at the six-month follow-up, compared to the approximately thirty-five percent success rate of varenicline, the most effective currently approved pharmacotherapy for smoking cessation. The sample was small and the study was not randomized, and it would be scientifically premature to treat these results as definitive. But the effect size is large enough, and the convergence with results from trials for depression, anxiety, PTSD, alcohol use disorder, and end-of-life existential distress is strong enough, to demand serious attention. Something remarkable is happening in these studies, and the effort to understand it at the level of the brain's electrical organization is one of the most illuminating frontiers in contemporary neuroscience.

From Sacred Medicine to Schedule One

Psilocybin, the active compound in more than two hundred species of mushrooms found on every inhabited continent, has been used by human beings for purposes of healing, spiritual exploration, and cultural ritual for at least five thousand years, and possibly much longer. The Aztec civilization of central Mexico used psilocybin mushrooms, which they called teonanacatl, the flesh of the gods, in religious ceremonies under the supervision of trained specialists. Shamanic traditions throughout Mesoamerica and parts of South America and Africa incorporated psilocybin-containing plants into healing rituals whose structure, involving preparation, ceremony, a guide or therapist present throughout the experience, and post-session integration of what occurred, bears a striking resemblance to the structure of modern clinical trials.

LSD, the synthetic compound that became the cultural symbol of the 1960s psychedelic movement, was first synthesized by the Swiss chemist Albert Hofmann at the Sandoz pharmaceutical company in 1938 and its psychological effects were discovered accidentally in 1943 when Hofmann absorbed a small amount through his fingertips in the laboratory. The subsequent twenty years saw an explosion of psychiatric research with LSD and, from 1958 onward, with psilocybin, which Hofmann also isolated and synthesized. More than a thousand clinical papers were published on psychedelic therapy between 1950 and 1965, reporting promising results in alcoholism, anxiety, depression, obsessive-compulsive disorder, and existential distress in terminal illness. The work was imperfect by modern standards, often lacking control groups and using outcome measures that would not satisfy a contemporary regulatory submission. But it represented a genuine and substantial body of evidence that something clinically significant was happening.

Then, in 1970, the United States Controlled Substances Act classified psilocybin and LSD as Schedule One drugs, a classification reserved for substances with high abuse potential and no accepted medical use, placing them in the same regulatory category as heroin and making clinical research involving them practically impossible for three decades. The scheduling was a political response to the cultural disruption of the 1960s rather than a scientific assessment of the evidence, as many of the researchers whose work was ended by it noted at the time with varying degrees of diplomatic restraint. The consequence was a thirty-year gap in the scientific investigation of compounds that the preliminary evidence suggested were among the most powerful tools for producing lasting changes in psychological functioning that pharmacology had ever encountered.

The Electrical Picture of Psilocybin

When the research renaissance began in the early 2000s, initially at Johns Hopkins under Roland Griffiths and at New York University under Stephen Ross, and subsequently at Imperial College London under Robin Carhart-Harris, one of the most important new capabilities that researchers brought to the investigation of psychedelics was modern neuroimaging. For the first time, it became possible to observe what psilocybin and LSD actually do to the brain's functional organization in real time, and the results were surprising in ways that changed how researchers thought about both the drugs and the brain.

The most striking finding from the first neuroimaging studies of psilocybin was that the drug decreased activity in the brain's key hubs of integrated information processing, rather than increasing it. The default mode network, whose chronic hyperactivity is associated with depression and rumination and whose activity corresponds to the narrative self, the brain's ongoing story of who you are and what matters to you, was dramatically suppressed by psilocybin. The thalamus, the brain's sensory relay and one of its key integration hubs, showed reduced activity. And the pattern of functional connectivity across the brain, the specific correlations between the activity of different regions that define the brain's normal organizational structure, was radically altered: regions that do not normally communicate directly with each other began to show correlated activity, while the normal boundaries between the brain's functional networks became less distinct.

Carhart-Harris and his colleagues at Imperial College described this altered state of brain organization using the concept of entropy: the psilocybin brain is a higher-entropy brain, one in which the normal constraints on which regions communicate with which are loosened, allowing a wider range of brain states and a richer, less predictable pattern of neural dynamics. The normal brain operates in a relatively low-entropy state, with activity strongly constrained by the habitual connectivity patterns that efficient information processing requires. The psilocybin brain temporarily moves into a higher-entropy regime in which those constraints are relaxed, connections that are normally suppressed become active, and the range of possible brain states expands dramatically.

This shift in entropy is, Carhart-Harris has argued, the neural mechanism that underlies both the phenomenology of the psychedelic experience, the dissolution of the normal boundaries of self and world, the extraordinary vividness and novelty of perception, the sense of seeing familiar things for the first time, and the therapeutic effects of psilocybin, the loosening of the rigid, habit-bound patterns of thought and feeling that maintain depression, addiction, and anxiety. The depressed brain, in this model, is a pathologically low-entropy brain: one whose activity is constrained into narrow, self-reinforcing patterns of negative rumination from which it cannot easily escape. The psilocybin experience temporarily disrupts these patterns, and the disruption, if the set and setting and integration are appropriately managed, can allow new patterns to form that are more adaptive and more consistent with the person's values and their sense of their own possibilities.

The Clinical Evidence

The clinical evidence for psilocybin-assisted therapy has accumulated rapidly in the past decade. In 2016, two simultaneously published trials in the Journal of Psychopharmacology, one from Johns Hopkins and one from NYU, reported the results of psilocybin-assisted therapy for anxiety and depression in patients with life-threatening cancer diagnoses. Both trials found dramatic reductions in anxiety and depression following a single session of psilocybin-assisted therapy, with effect sizes substantially larger than those typically seen with conventional pharmacotherapy, and with improvements that were sustained at follow-up periods of six months and, in the Johns Hopkins trial, up to four and a half years. A substantial proportion of participants described the psilocybin experience as among the most meaningful experiences of their lives, and the mystical-type quality of the experience, the sense of unity, sacredness, and noetic quality that characterizes the full psilocybin response, was among the best predictors of therapeutic outcome.

For depression specifically, Robin Carhart-Harris and his colleagues at Imperial College and subsequently at UCSF published a series of trials showing that psilocybin-assisted therapy produced rapid and substantial antidepressant effects in patients with treatment-resistant depression, comparable in magnitude to the effects of conventional antidepressant medications but occurring after two sessions rather than weeks or months of daily medication. A larger randomized trial published in the New England Journal of Medicine in 2021 compared psilocybin to escitalopram, a widely used SSRI, in patients with moderate to severe depression, finding that both treatments produced significant improvements in depression scores but with different profiles: escitalopram showed greater effects on some symptom measures while psilocybin showed greater effects on others, including measures of emotional processing and wellbeing that SSRIs characteristically do not improve.

For PTSD, the compound under investigation is MDMA rather than psilocybin. MDMA, which gained notoriety in the 1980s as the recreational drug Ecstasy, acts on serotonin, dopamine, and norepinephrine systems to produce a distinctive state of emotional openness, reduced fear, and enhanced empathy that researchers at the Multidisciplinary Association for Psychedelic Studies have proposed as therapeutically useful for the treatment of PTSD, specifically because it allows patients to revisit traumatic memories with reduced fear reactivity, facilitating the emotional processing of those memories in psychotherapy. Phase 3 clinical trials of MDMA-assisted therapy for PTSD, published in Nature Medicine in 2021 and 2023, found that sixty-seven percent of participants who received MDMA-assisted therapy no longer met diagnostic criteria for PTSD at the end of treatment, compared to thirty-two percent of those who received therapy without MDMA. These results led the FDA to grant MDMA Breakthrough Therapy designation, a regulatory category reserved for treatments that show substantial improvement over available therapy for serious conditions.

The Integration Problem

The evidence for psychedelic therapy's clinical efficacy, while remarkable in its effect sizes, raises a scientific question that is both practically important and theoretically fascinating: why does an experience that lasts for a few hours produce clinical benefits that persist for months or years? The conventional model of psychiatric pharmacotherapy involves daily medication that maintains a continuous biochemical effect. The psychedelic model involves a brief, intense intervention that somehow produces lasting change in the brain's functional organization. Understanding the mechanism of this persistence is one of the most active areas of research in psychedelic neuroscience.

The leading hypothesis, supported by a combination of animal studies and human neuroimaging data, is that psilocybin and related compounds promote neuroplasticity: the growth of new synaptic connections in the prefrontal cortex and other brain regions that have been implicated in the therapeutic response. Animal studies have shown that psilocybin produces rapid and substantial increases in dendritic spine density in the prefrontal cortex, the structural substrate of new synaptic connections, within twenty-four hours of administration, and that these structural changes persist for at least a month. If these findings translate to humans, they suggest that the therapeutic window opened by psilocybin is one in which the brain is in a state of elevated neuroplasticity, more readily forming new connections and new patterns of activity, and that the quality of the psychological work done during this window, the nature of the therapeutic session and the integration of what occurred afterward, determines what new patterns are established and how lasting they are.

This is why the set and setting, the psychological context in which the psychedelic experience occurs, matter so much in determining the outcome of psychedelic therapy. The drug opens a window. What comes through that window, what the person encounters in the experience and how they make sense of it afterward with the help of an experienced therapist, determines what the opened window is for. The electrical rewiring that psilocybin initiates is not predetermined.

It is shaped by everything the person brings to the experience and everything they do with it in its aftermath. The biology is profound. The psychology is equally profound. Neither is sufficient without the other.

PART FIVE

THE ELECTRIC FUTURE

Brain-Computer Interfaces, Neural Rights, and What It Means to Be Human

Part Introduction

The Edge of What It Means to Be Human

The chapters of Parts One through Four have described the brain's electrical life as it currently exists and as medicine is learning to intervene in it. Part Five moves to the edge of what is currently happening and what is beginning to happen, to the technologies and the questions that are arriving faster than our institutions and our philosophy and perhaps our imagination are ready for. The two chapters of Part Five are the most speculative in this book, not in the sense of being unsupported by evidence but in the sense of being about a future that is already partly present and whose full implications have not yet been worked out by anyone.

They are also, in many respects, the most important chapters in the book, because they are about a set of developments that will affect everyone who has a brain, which is everyone, and that are raising questions about identity, freedom, equality, and the nature of the self that will need to be answered by the whole of society rather than by any single discipline or institution. The scientists who are developing these technologies are, in the main, acutely aware of the ethical dimensions of their work and deeply engaged with the questions it raises. But scientific awareness of ethical complexity is not the same as political and legal response to it, and the response is currently lagging badly behind the pace of the science.

Part Five asks two questions. Chapter Fourteen asks what happens when we merge human minds with machines, not metaphorically but literally, at the level of neurons and electrodes and data. Chapter Fifteen asks who owns the resulting information, who has the right to alter or enhance the electric self, and what protections human beings will need in a world where the most intimate facts about their inner lives are accessible to anyone with the right technology and the right access. These are the most urgent questions that neuroscience is raising for the twenty-first century. They deserve the most careful and the most honest attention we can give them.

CHAPTER FOURTEEN

The Merged Mind

Neuralink, Brain-Computer Interfaces, and the Future of Human Intelligence

ON JANUARY 29, 2024, a twenty-nine-year-old man named Noland Arbaugh became the first human being to receive a Neuralink brain-computer interface. Arbaugh had been paralyzed from the shoulders down since 2016, when a diving accident left him with a spinal cord injury at the C4-C5 level. He had watched, over the years since his injury, as the BrainGate research program demonstrated increasingly impressive neural control of computers and robotic devices in a series of trial participants, and he had applied to Neuralink's PRIME study, the company's first human clinical trial, when it opened for applications in 2023. He was accepted. On the morning of the surgery, a robotic surgical system guided by real-time MRI imaging inserted 64 flexible electrode threads, each thinner than a human hair, into the motor cortex of his brain, with the threads positioned to record from approximately 1,024 neurons simultaneously, ten times more than any previous clinical brain-computer interface.

Within weeks of the implantation, Arbaugh was controlling a computer cursor with his thoughts, moving it across the screen by imagining the movement of his paralyzed hand. Within months, he was using the system to play online chess, to browse the internet, to play video games, and to engage in the various activities of digital life that had become inaccessible to him after his injury. He gave interviews, appeared on podcasts, and became, in the absence of any competing narrative from the secretive company that had implanted his device, the public face of a technology that was simultaneously the most advanced brain-computer interface ever deployed in a human and the most visible demonstration that the science of neural decoding had crossed a threshold from research curiosity to practical clinical tool.

Elon Musk, whose company Neuralink had been developing its device since its founding in 2016, had described the long-term vision of the technology not as a medical device for paralyzed patients but as a general augmentation platform: a means of merging human intelligence with artificial intelligence to create cognitive capabilities that neither could achieve alone. The medical application was, in Musk's framing, a necessary first step toward regulatory approval and public acceptance, a compassionate entry point into a technology whose ultimate purpose was far more ambitious than restoring function to people who had lost it. Whether one finds this vision exciting or alarming, or both, depends on a great deal about what one believes about the relationship between the brain and the self, between technology and humanity, and between the possibilities and the risks of accelerating change in the most intimate possible domain.

The Landscape of Neural Augmentation

Neuralink is the most publicly visible company in the brain-computer interface space, but it is not the only one, nor in terms of current clinical capability, necessarily the most advanced. The BrainGate consortium, which has been conducting human trials since 2004 and has published the most rigorous and extensive clinical data of any group in the field, uses devices manufactured by Blackrock Neurotech and has demonstrated capabilities that in some respects exceed those Neuralink has reported. Synchron, an Australian-American company, has developed a BCI device called the Stentrode that is delivered endovascularly, through the blood vessels, without open-brain surgery: a wire mesh device that is threaded into a blood vessel near the motor cortex and deploys to record neural signals through the vessel wall. Several Stentrode patients have demonstrated typing speeds of around fifteen words per minute using the system, a more modest performance than the best invasive systems but achieved with a substantially less invasive procedure.

Kernel, another venture-backed company, is pursuing a different approach entirely: instead of implanting electrodes to record neural activity directly, Kernel is developing advanced non-invasive neuroimaging technology capable of recording brain activity with greater spatial and temporal resolution than current methods. The company's helmet-based devices use magnetoencephalography, the measurement of the tiny magnetic fields generated by neural currents, with room-temperature sensors that do not require the expensive liquid helium cooling of conventional MEG systems, and functional near-infrared spectroscopy, an optical brain imaging technique that measures changes in blood oxygenation. The resulting system provides brain measurements of unprecedented quality for a wearable, non-invasive device, and could in principle give a large population of user's access to a form of neural interface without any of the surgical risks of implanted systems.

Beyond these companies, an entire ecosystem of research programs, startups, and established medical device manufacturers is working on various aspects of the brain-computer interface problem: better electrode materials that cause less tissue damage and maintain recording quality for longer periods, wireless data transmission systems that can handle the enormous data rates generated by large electrode arrays without the heating and power consumption problems that constrain current systems, machine learning algorithms that can extract more useful information from noisy neural recordings, and implantable devices that can not only record neural activity but also deliver precisely targeted stimulation in response to what they record, closing the loop between reading and writing the brain's electrical state.

The Enhancement Question

The medical applications of brain-computer interfaces, restoring motor control to paralyzed patients, restoring communication to those who have lost the ability to speak, treating neurological and psychiatric conditions with precisely targeted stimulation, have a clear ethical justification that most people find readily intelligible. Restoring a capacity that a disease or injury has taken is different in kind from adding a capacity that nature has not provided, and the distinction between restoration and enhancement maps onto a moral intuition that is widely shared across different ethical frameworks.

But the distinction is not as clear in practice as it appears in principle. When a cochlear implant restores hearing to a person who was born profoundly deaf, is this restoration or enhancement? The person never had the hearing that is being provided; the implant is not restoring something lost but providing something never possessed, and it is providing it in a form that is different from natural hearing in important ways that some members of the Deaf community regard as a significant loss of cultural identity rather than a gain of function. When a DBS device corrects a mood disorder and makes a person more emotionally stable and productive than they were before the disorder, is the improved state a restoration of their authentic self or an enhancement beyond it? The questions do not have obvious answers, and the answers matter greatly for how we regulate these technologies.

For enhancement specifically, the ethical concerns are multiple and serious. If a neural augmentation device can make a person smarter, more attentive, better at learning, more emotionally regulated, and more productive than any unaugmented human being, who will have access to it? The history of transformative technology suggests that the answer, at least initially, is those who can pay for it. And the consequent stratification of human cognitive capability, the creation of a class of cognitively augmented individuals who can outcompete all unaugmented competitors in every domain that matters for professional and economic success, raises concerns about inequality that are at least as serious as those raised by other forms of inequitable access to powerful technology, and possibly more so, because cognitive advantage compounds over time and across generations in ways that other forms of advantage do not.

The counter-argument, made by some advocates of neural augmentation, is that cognitive enhancement technology will follow the same trajectory as other transformative technologies: initially expensive and accessible only to the wealthy, then rapidly commoditizing and becoming accessible to a much broader population, eventually becoming a standard part of the technological infrastructure of everyday life in the way that smartphones have. The optimistic view notes that smartphones, which were initially luxury items affordable only to affluent early adopters, are now used by billions of people across the full range of economic circumstances, and have provided access to information, communication, and economic opportunity to populations that would otherwise have had none of these things.

Whether this optimistic trajectory is plausible for neural augmentation technology specifically depends on many factors that are currently impossible to predict with confidence: the pace of technological development, the regulatory environment, the economics of the healthcare system, and above all the question of safety. Neural augmentation devices that are implanted in the brain carry risks, both from the surgical procedure and from the long-term presence of foreign material in brain tissue, that are different in kind from the risks of carrying a smartphone. The risk-benefit calculation for a device that provides convenience is different from one that provides medical necessity, and the social pressure on individuals to adopt augmenting technologies could be powerful in ways that compromise the voluntary character of the choice.

The Identity at the Merge

There is a question that sits behind all of the practical and ethical questions about brain-computer interfaces, and it is the question that this chapter has been approaching throughout. When a person's cognitive capabilities are augmented by a device that is physically part of their brain, recording its activity and stimulating it in response, are the resulting thoughts and capabilities their own? If an augmented person has an insight that they could not have had without the device, does the insight belong to them? If a neural interface enhances a person's memory by recording and replaying information that their biological memory could not retain, are those enhanced memories authentic parts of their identity? If a device makes a person more empathetic or more patient or more focused than their unaugmented neurobiology would allow, are these enhanced qualities genuinely theirs?

These are not merely philosophical puzzles. They have practical legal consequences: for intellectual property law, which requires the identification of inventors; for criminal law, which requires the identification of culpable agents; for contracts, which require the identification of competent decision-makers; and for democracy, which requires the identification of autonomous voters. They also have deeply personal consequences for the people who will use these technologies and who will have to find a way to think about their own identity in relation to the device that is helping them think.

My own view, which I offer with the caveat that it is a view rather than an established fact, is that the identity question, asked honestly, reveals that the concern it expresses is not new. Human beings have always used external tools to augment their cognitive capabilities: language, writing, mathematics, books, libraries, the internet, the smartphone. Each of these technologies has extended human cognitive reach beyond what biology alone would allow. The thought you have using language is no less authentically your thought for having been enabled by a cultural technology that was not invented by biology. The memory stored in a notebook is no less genuinely your memory for being encoded in ink rather than in synaptic weights. What brain-computer interfaces represent is not a new kind of cognitive augmentation but an extension of the existing continuum of cognitive augmentation to a more intimate physical scale. Whether this difference of scale constitutes a difference of kind, whether the crossing of the skull boundary changes the nature of the relationship between the tool and the self in ways that matter morally, is the question. I am not certain. I am not sure anyone is. But I think the question deserves more careful and more urgent attention than it is currently receiving in most of the public discussions of this technology.

CHAPTER FIFTEEN

Who Owns Your Electric Self?

Neurorights, Mental Privacy, and the Freedom of Thought in the Age of Neural Technology

"The brain is the last and greatest biological frontier. The most complex thing we have yet discovered in our universe. It contains hundreds of billions of cells interlinked through trillions of connections. The brain boggles the mind."

— James Watson, 1992

"If neurotechnology companies can decode and manipulate what goes on in our minds, those of us alive now are the last generation with mental privacy."

— Rafael Yuste, 2023

BEGIN WITH A THOUGHT experiment. Imagine that tomorrow, a company announces the commercial release of a wearable EEG headset, slightly more sophisticated than the consumer devices currently available, that can detect with reasonable accuracy whether the person wearing it is experiencing positive or negative emotional responses to stimuli presented to them. The device is marketed as a wellness tool: wear it while you watch Netflix, and it will curate your viewing choices to maximize positive emotional response. Wear it while you shop, and it will highlight products your brain responds to most favorably. Wear it during a conversation, and it will later provide you with an analysis of which topics made you anxious and which made you comfortable.

Now imagine that the terms of service of this device, buried in a document that forty thousand words long and that ninety-eight percent of users will not read, include a clause stating that the neural data collected by the device is the property of the company, that the company may share it with third parties including advertisers and employers and insurers, and that the data may be retained indefinitely and used to train the machine learning algorithms that will be incorporated into the next generation of the product. The company complies with all current privacy laws in every jurisdiction in which it operates, because no current privacy law specifically addresses neural data. The device is legal. The data collection is legal. The sale of the data is legal. The use of the data to make inferences about a person's emotional states, their susceptibilities, their health, their political views, and their likely behavior in a wide range of future situations is legal.

This is not a fantasy. It describes a situation that is essentially already present with current commercial EEG technology, and that will become considerably more serious as the decoding algorithms improve and as the devices become more capable. The thought experiment is not designed to produce alarm for its own sake. It is designed to make concrete a problem that is currently discussed in abstract terms by neuroscientists, ethicists, and legal scholars but has not yet entered the mainstream political conversation with the urgency it deserves: the problem of neural privacy, and the broader question of who has the right to access, use, alter, and profit from the information contained in the electric brain.

The Five Neurorights

The neurorights framework, developed most systematically by Rafael Yuste and his colleagues at the NeuroRights Foundation at Columbia University, identifies five specific rights that existing legal frameworks fail to protect and that the advance of neurotechnology makes urgently necessary. Understanding these five rights, and why each of them matters, is a good way to understand both the nature of the threat and the shape of the response that is needed.

The first neurightis mental privacy: the right to keep the contents of one's mind private from external observation without one's consent. This is an extension of the existing right to privacy into the neural domain, and it is the most immediately threatened of the five rights by current and near-term neurotechnology. Neural data, as the chapters on BCI and neural decoding have established, contains information about a person's emotional states, cognitive processes, intentions, and in some cases specific thoughts, and current law in most jurisdictions provides no specific protection for this information once it has been collected by a device that the person is voluntarily wearing or has voluntarily implanted.

The second neuroright is mental integrity: the right to protection from unauthorized manipulation of one's neural states by external technology. This right becomes relevant not only for future enhancement technologies but for existing therapeutic technologies whose off-label or non-consensual use could produce behavioral changes in individuals who have not agreed to be altered. A stimulation device capable of changing a person's emotional state or decision-making patterns is a device capable of manipulation if used without full informed consent, and the governance frameworks for ensuring that consent is genuine and ongoing rather than a one-time signature on a lengthy term of service document do not yet exist.

The third neuroright is psychological continuity: the right to protection from interventions that would produce such a radical change in one's personality or sense of self as to constitute a fundamental disruption of personal identity across time. This right is relevant for enhancement technologies that could produce profound changes in personality, cognition, or emotional character, and it raises the same philosophical questions about identity that the chapter on DBS began to address. If a neural intervention changes who you are in a sufficiently fundamental way, was the person who agreed to the intervention the same person who lives with its consequences?

The fourth neuroright is cognitive liberty: the right to alter one's own mental states, including through the use of technology or pharmacology, and the corresponding right to refuse such alteration if others seek to impose it. Cognitive liberty is a two-edged principle: it protects the freedom to enhance or alter one's own mind voluntarily, which matters for people who might be prevented from using beneficial technologies by overly restrictive regulation, and it protects the freedom from having one's mind altered without consent, which matters for individuals who might be coerced into neural interventions by employers, insurers, governments, or social pressure.

The fifth neuroright is equitable access to mental augmentation: the principle that if neural enhancement technologies become available, they should not create or deepen cognitive inequality between those who can afford them and those who cannot. This is perhaps the most politically complex of the five rights, because it requires some positive obligation on the part of governments or markets to ensure broad access to technologies that may be expensive to develop and deploy, and the mechanisms for fulfilling this obligation are not obvious. But the alternative, a world in which some humans are cognitively augmented and others are not, and in which the augmented consistently outcompete the unaugmented in every domain that matters for economic and social opportunity, is one that virtually every framework of justice would regard as unacceptable.

Chile and the Constitutional Moment

In 2021, Chile became the first country in the world to enshrine neural rights in its constitution. The amendment, passed unanimously by the Chilean Senate in September of that year, added neural data to the list of protected personal data under Chilean law, established that the development and use of neurotechnology must respect mental integrity and free will, and placed the regulation of neurotechnology within the framework of existing human rights law. It was a modest but historically significant step: the first time that any national government had recognized, at the constitutional level, that the advance of neurotechnology created a new category of rights that required explicit legal protection.

The Chilean amendment was followed by the passage of neurorights legislation in Chile's legislature in 2021, creating specific regulatory provisions for the commercial use of neural data, and by the introduction of similar legislation in several other jurisdictions. The province of Ontario in Canada proposed neurorights legislation in 2022. The State of Colorado in the United States enacted a law in 2024 protecting neural data as a category of sensitive personal information under its existing privacy law. The European Union began deliberations on how to incorporate neural data protections into its General Data Protection Regulation framework. The United Nations Educational, Scientific and Cultural Organization issued guidance on the ethics of neurotechnology. The pace of legal and regulatory development, while still lagging the pace of the science, has accelerated significantly since Yuste and his colleagues began their advocacy campaign in the late 2010s.

Yuste himself is cautiously optimistic but not complacent. The legislative victories of the early 2020s, he has argued in his public lectures, are meaningful but insufficient: they create symbolic recognition of the problem without yet providing the robust, enforceable regulatory framework that would be needed to actually protect neural privacy and mental integrity in a world where the technology of neural reading and writing is advancing as rapidly as it is. The challenge is similar to the challenge of regulating the internet in the 1990s and early 2000s: the technology moved faster than the regulation, the regulatory frameworks were designed for a world that the technology had already superseded, and the consequences of that lag are still being addressed two decades later. With neurotechnology, the stakes are higher and the window for getting the regulation right before the technology outruns it is narrower.

The Dual-Use Problem

One of the most persistent challenges in developing governance frameworks for neurotechnology is the dual-use character of most neural technologies: the same device that can restore communication to a locked-in patient can potentially be used to decode the thoughts of an unwilling subject; the same stimulation technology that can relieve depression can potentially be used to alter mood and behavior without consent; the same neural decoding algorithms that can reconstruct language for medical purposes can potentially be used to reconstruct the private thoughts of anyone whose brain activity can be recorded.

The dual-use problem is familiar from other domains of powerful technology. Nuclear reactors produce electricity and nuclear weapons. Genetic engineering produces life-saving medicines and potentially misused biological agents. The internet enables unprecedented information sharing and unprecedented surveillance and manipulation. In each of these cases, society has developed, with varying degrees of success, governance frameworks that attempt to preserve the beneficial uses while limiting the harmful ones. The challenge with neurotechnology is that the beneficial and harmful uses are more intimately connected, more technically similar, and more difficult to distinguish at the level of the device itself than in most other dual-use domains.

A neural decoding system designed to help a locked-in patient communicate is, technically, a neural surveillance system that does not require the subject's cooperation to function, provided that the subject's brain can be recorded. The difference between the therapeutic and the surveillance applications is not in the technology but in the context: the consent of the subject, the oversight of the use, the legal framework governing the application. Getting the governance right therefore requires not only technical standards for neural devices but also robust consent frameworks, use-limitation requirements, data minimization principles, and enforcement mechanisms that can actually constrain the behavior of powerful actors who might prefer to use neural technology in ways that serve their interests rather than those of the individuals whose brains they are reading.

The Deeper Question

Behind all of the specific legal and policy questions about neural rights, there is a deeper question that this chapter, and in a sense this entire book, has been building toward. The question is this: what is the relationship between the electric brain and the person who has it? Is the brain the person, in which case any technology that accesses or alters the brain is accessing or altering the person in the most fundamental possible sense? Or is the person something that the brain supports and enables but that is not reducible to the brain, in which case the relationship between neural technology and personal identity is somewhat less direct?

The answer to this question, or rather the position one takes on it given that it does not have a settled answer, has profound consequences for how one thinks about neural rights and neural governance. If the brain is the person, then reading a person's neural data is reading the person, and the right to mental privacy is as fundamental as any right can be. If the person is more than the brain, then neural data, however intimate, is still in some sense at one remove from the person, and the right to privacy of neural data, while important, is not qualitatively different from the right to privacy of other forms of sensitive personal information.

My own view is that the brain is, in the most important sense, the person: that everything that makes you specifically you, your memories, your values, your personality, your characteristic ways of experiencing and interpreting and responding to the world, is encoded in the physical structure and electrical activity of your brain, and that there is no additional self that exists independently of this biological substrate.

But I also believe that this view, properly understood, does not reduce the person to a mere mechanism. It does the opposite: it makes the brain, and therefore the person, the most precious and the most worthy of protection physical object in the known universe. A universe that understands itself through consciousness is a universe of astonishing improbability, and the individual biological engines of that understanding, the 86 billion neurons and 100 trillion synapses of each human brain, are the most valuable things in it.

If you accept this, then the governance of neurotechnology is not merely a legal and regulatory problem. It is a civilizational priority. It is the question of whether the most intimate and the most irreplaceable thing that each person has, the specific, unique, irreproducible electrical pattern that is their mind, will be protected in the age of technology that can read it, write it, augment it, and market it. The answer to that question is not determined by the science. It is determined by the choices that society makes, the values it chooses to prioritize, the rights it chooses to enshrine, and the institutions it chooses to empower to enforce those rights. Those choices are being made now, in the early decades of the twenty-first century, at a moment when the science has advanced further than most people realize and the governance has not kept pace. They are among the most important choices that this generation of human beings will make, for themselves and for the generations that will inherit whatever world those choices create.

EPILOGUE

The Light That Knows It Shines

WE BEGAN THIS BOOK in a laboratory in Bologna in the autumn of 1780, with a dead frog and a copper wire and the electrical shock of an observation that nobody was quite prepared for. We end it in 2026, in a world where paralyzed patients play chess with their thoughts, where the contents of working memory can be reconstructed from brain scans, where implanted electrodes restore speech to people who have been silent for eighteen years, where psychedelic molecules open windows of neuroplasticity that may offer relief to millions of people for whom conventional psychiatry has run out of answers. The distance between Galvani's frog and Noland Arbaugh's Neuralink implant is 244 years and represents one of the most extraordinary arcs of scientific progress in the history of any field.

But the distance between Galvani's central question and the question that closes this book is much smaller than 244 years would suggest. Galvani wanted to know what electricity had to do with life. We still want to know. We have answered vast amounts of the subsidiary question, we know the mechanisms of the action potential, the architecture of the memory trace, the circuit organization of emotion, the electrical signatures of sleep, the specific dysregulations that drive psychiatric and neurological disease, the ways in which technology can read and write the brain's electrical language. We know, in a sense that would have been entirely beyond Galvani's imagination, an extraordinary amount about how the brain's electricity works. We still do not know, in any satisfying sense, why it feels like anything at all.

This is the Hard Problem, and it remains hard. The theories of consciousness described in Chapter Seven, Integrated Information Theory, Global Workspace Theory, the various quantum approaches, are genuine attempts by serious scientists and philosophers to bridge the explanatory gap between mechanism and experience. None has yet succeeded in a way that has convinced the field. The gap remains. And I want to end this book by arguing, against the impatience that sometimes characterizes scientific discourse, that the persistence of the gap is not a failure of neuroscience. It is a signal about the depth of the question.

Consider what consciousness is, from the perspective of the universe as a whole. The universe is approximately 13.8 billion years old. For the first several billion of those years, it was entirely without experience: matter and energy interacting according to physical laws, but with no inner perspective on that interaction, no point from which the universe was experienced from the inside. Then, through a process of evolution whose details are still being worked out, biological organisms developed nervous systems. And at some point in that process, for the first time in the history of the universe, the lights came on. The matter of the universe began to experience itself. There was, for the first time, something it was like to be something. The universe had, in the most literal possible sense, woken up.

That waking up happened in brains. Specifically, in the electrical activity of brains. The lights that came on are the action potentials and the oscillations and the synaptic events that this book has described. The thing that woke up is what you are, and what every other conscious being is: a particular electrical pattern in biological tissue, unique and irreproducible and temporary, experiencing itself and the universe from a specific vantage point that has never existed before and will never exist again. This is not a diminishment. This is the most remarkable thing in the known universe. The matter of the universe is looking at itself and wondering what it is. You are the universe doing that. Right now, in this moment, reading these words with your brain's visual cortex and your language areas and your working memory and your prefrontal cortex and your emotional systems all engaged simultaneously in the specific electrical dance that is reading and understanding and responding, you are the universe's most sophisticated instrument of self-reflection.

The fact that we do not yet know how electricity becomes experience does not make that experience less real or less extraordinary. It makes it more so: it means that what you are is, at its most fundamental level, still a mystery. Not a mystery of the kind that suggests ignorance or incompetence. A mystery of the kind that suggests depth: that what you are exceeds, in its richness and its irreducibility, every description that has yet been given of it. The Hard Problem is hard because what it is pointing at is genuinely extraordinary: the fact that the universe not only exists but is experienced, and that this experience is, at every moment and for every conscious being, as vivid and as present and as undeniably real as anything that physics or chemistry or neuroscience has ever measured.

There is a phrase that I have been turning over for the duration of writing this book, trying to find the right place for it, and I think this is where it belongs. The light that knows it shines. It refers to consciousness, to the fact that experience is not only present but self-aware: that you do not merely have experiences, you know that you are having them. You are aware of your awareness. The light does not merely illuminate. It knows that it illuminates. This recursive quality of consciousness, the way that experience loops back on itself to become the subject of its own awareness, is one of the features of the Hard Problem that makes it most resistant to physical explanation, because physical systems, as normally described, do not have this quality. A photon does not know that it is light. A neuron does not know that it is firing. But you know that you are experiencing. And that knowing, that self-luminous quality of consciousness, is what makes you something that the physical description of your brain, however complete, somehow cannot fully capture.

I do not know how to explain this. I have read everything that the most serious scientists and philosophers have written about it, and I do not think anyone else knows how to explain it either, though many people have valuable and important things to say about its edges and its correlates and its consequences. What I do know is that it is real, and that it is yours, and that the fact that it is not yet fully explained by science is the most appropriate possible condition for the book's final pages. Because the most honest thing I can say, after everything that has been described in these three volumes, is this: the brain is astonishing. We understand it incompletely and increasingly. It generates, through mechanisms that are real and measurable and beautiful in their complexity, the experience of being you. And the being of you, the specific, vivid, unrepeatable electrical storm that is your consciousness, remains the most important unsolved mystery in all of human knowledge.

It is, as Christof Koch said, the most important thing in the universe.

Protect it accordingly.

THE END

The Electric Self

Mohammad Umar

India 2026

ACKNOWLEDGEMENTS

A book of this scope is not written in isolation, and the debts accumulated in its writing are considerable. I am grateful, first and most fundamentally, to the scientists whose work this book describes: the researchers past and present whose decades of patient, rigorous investigation has built the edifice of neuroscience that this book attempts to make accessible to a wider audience. I have tried to represent their work accurately and fairly, and any errors of fact or interpretation are entirely my own.

The history of the science in Part One draws heavily on the histories written by Stanley Finger, particularly his comprehensive Origins of Neuroscience, and on Max Bennett and Peter Hacker's Philosophical Foundations of Neuroscience, which remains the most rigorous account of the conceptual underpinnings of the field. The neuroscience of memory in Chapter Four owes an enormous intellectual debt to the work of Eric Kandel, whose autobiography In Search of Memory is among the finest accounts of a scientific career I have ever read, and to the late Brenda Milner, whose decades of work with Henry Molaison laid the foundations of everything we know about the hippocampus and declarative memory. The emotional neuroscience of Chapter Five draws primarily on the published work of Joseph LeDoux, whose The Emotional Brain and Anxious are essential reading for anyone interested in the biology of fear and emotion.

The sleep science of Chapter Six is indebted to Matthew Walker, whose Why We Sleep brought the science of sleep to a broad popular audience with admirable clarity and passion, and to the primary research of Maiken Nedergaard, whose discovery of the glymphatic system is one of the most important findings in neuroscience of the past decade. Chapter Seven's treatment of the Hard Problem of consciousness relies primarily on the philosophical work of David Chalmers, whose The Conscious Mind remains the clearest and most rigorous statement of the problem, and on the empirical research programs of Giulio Tononi and Stanislas Dehaene, whose theoretical frameworks continue to organize the empirical investigation of neural correlates of consciousness.

The clinical chapters of Part Three drew on conversations with clinicians and researchers too numerous to list individually, and on the patient narratives that give those chapters their human center. I am grateful to the patients who have shared their stories in published memoirs and interviews, and to the researchers who have made those stories central to their scientific accounts. The chapters of Parts Four and Five rely on the primary literature of neuromodulation and brain-computer interfaces, and on the public advocacy of Rafael Yuste and the NeuroRights Foundation, whose work to make the governance of neurotechnology a mainstream policy concern is among the most important scientific-advocacy efforts of the current decade.

Finally, I am grateful to my readers, whose willingness to spend time with a difficult and important subject is both the purpose and the reward of this work.

Mohammad Umar

GLOSSARY OF KEY TERMS

ACTION POTENTIAL: THE fundamental electrical signal of the neuron. A rapid, all-or-nothing change in membrane voltage that propagates along the axon from the cell body to the axon terminals. Generated by the sequential opening of voltage-gated sodium and potassium channels in the axonal membrane.

Amygdala: An almond-shaped structure in the medial temporal lobe, present in both hemispheres. Central to the processing of emotional information, particularly fear responses and the emotional modulation of memory formation. Receives rapid, crude sensory input via the thalamus (the low road) and more detailed cortical input (the high road).

Axon: The long projection of a neuron that carries electrical signals away from the cell body to its targets. May be wrapped in myelin to increase conduction velocity. Some axons extend over a metre in length in large animals.

Brain-computer interface (BCI): Any device or system that creates a direct communication pathway between the brain and an external computer or device. BCIs may be invasive, with electrodes implanted in the brain, or non-invasive, relying on external recordings of neural activity.

Deep brain stimulation (DBS): A neurosurgical intervention in which electrodes are implanted in specific brain targets and connected to an implantable pulse generator that delivers continuous electrical stimulation. Approved for Parkinson's disease, essential tremor, dystonia, obsessive-compulsive disorder, and epilepsy; under investigation for depression, Alzheimer's disease, and other conditions.

Default mode network (DMN): A network of brain regions, including the medial prefrontal cortex, posterior cingulate cortex, and angular gyrus, that is more active during rest than during externally directed tasks. Associated with self-referential thought, mind-wandering, episodic memory retrieval, and social cognition. Chronically hyperactive in major depression.

Dendrite: Branching projections of a neuron that receive incoming signals from other neurons at synaptic contacts and conduct them toward the cell body.

EEG (Electroencephalography): A technique for recording electrical activity from the brain using electrodes placed on the scalp. Measures the summed electrical activity of large populations of synchronously active neurons. Excellent temporal resolution (milliseconds) but limited spatial resolution.

Gamma oscillations: Brain waves in the 30 to 100 hertz frequency range associated with active sensory processing, cognitive engagement, and, in some theories, perceptual binding and conscious awareness. Disrupted in Alzheimer's disease, schizophrenia, and other conditions.

Glymphatic system: A brain-specific waste clearance system that uses channels surrounding blood vessels, lined with astrocytes, to flush cerebrospinal fluid through brain tissue during sleep. Removes metabolic waste products including amyloid-beta. Discovered in 2013 by Maiken Nedergaard and colleagues.

Hard Problem of consciousness: The philosophical problem, named by David Chalmers in 1995, of explaining why and how physical processes in the brain give rise to subjective experience: why there is something it is like to be conscious rather than information being processed in the dark without any inner experience.

Hippocampus: A curved structure in the medial temporal lobe essential for the formation of new declarative memories and for spatial navigation. Particularly vulnerable to damage in Alzheimer's disease and hypoxia. Named for its resemblance to a seahorse.

Long-term potentiation (LTP): A persistent strengthening of synaptic connections that is induced by repeated, simultaneous activation of pre- and postsynaptic neurons. Dependent on NMDA receptor activation and calcium influx. Widely regarded as the primary cellular mechanism of learning and memory.

Myelin: A fatty substance that forms an insulating sheath around axons, produced by Schwann cells in the peripheral nervous system and oligodendrocytes in the central nervous system. Dramatically increases the speed of action potential conduction by allowing saltatory conduction between nodes of Ranvier.

Neural decoding: The process of reconstructing information about a person's sensory experience, cognitive state, or behavioral intention from measurements of brain activity. Advanced by machine learning algorithms applied to fMRI, EEG, or intracortical electrode recordings.

Neuron: The fundamental cell type of the nervous system, specialized for the generation and transmission of electrical signals. Characterized by a cell body, dendrites that receive input from other neurons, and a single axon that transmits output to target cells.

Neurorights: A proposed category of human rights specifically addressing the protection of mental privacy, mental integrity, psychological continuity, cognitive liberty, and equitable access to neural augmentation in the context of advancing neurotechnology. First enshrined in constitutional law by Chile in 2021.

Neurotransmitter: A chemical messenger released from the presynaptic axon terminal in response to an action potential. Diffuses across the synaptic cleft and binds to receptor proteins on the postsynaptic membrane. Major neurotransmitters include glutamate (excitatory), GABA (inhibitory), dopamine, serotonin, norepinephrine, and acetylcholine.

Neuroplasticity: The ability of the brain to change its structure and function in response to experience, learning, injury, or intervention. Includes synaptic plasticity, changes in synaptic strength, and structural plasticity, changes in the physical connections between neurons.

Prefrontal cortex (PFC): The anterior portion of the frontal lobes, particularly expanded in humans relative to other primates. Essential for working memory, planning, impulse control, decision-making, and the regulatory control of emotion via descending connections to limbic structures.

Psilocybin: The active compound in more than two hundred species of fungi, producing altered states of consciousness including dissolution of normal boundaries of self and world, heightened sensory experience, and profound emotional and spiritual experiences. Currently under clinical investigation for treatment-resistant depression, PTSD, addiction, and end-of-life anxiety.

REM sleep: Rapid Eye Movement sleep. A stage of sleep characterized by rapid eye movements, muscle atonia, and brain activity resembling wakefulness. The stage in which most vivid dreaming occurs. Associated with emotional memory consolidation and creative problem-solving.

Subgenual anterior cingulate cortex (Area 25): A region at the junction of the prefrontal cortex and limbic system, chronically hyperactive in major depression and normalized by all effective antidepressant treatments. The primary target of deep brain stimulation for treatment-resistant depression.

Synapse: The specialized junction between two neurons where information is transmitted. Consists of the presynaptic axon terminal, the synaptic cleft of approximately 20 nanometres, and the postsynaptic membrane of the receiving neuron's dendrite or cell body.

Transcranial magnetic stimulation (TMS): A non-invasive brain stimulation technique that uses a changing magnetic field to induce electrical currents in cortical neurons beneath the stimulating coil. FDA-approved for major depression, OCD, and other conditions. Does not require anaesthesia or surgery.

ABOUT THE AUTHOR

MOHAMMAD UMAR IS A science writer and researcher based in Prayagraj, India. He is the author of The Hidden Universe Within: How Trillions of Microbes Control Your Mind, Body, and Destiny, which explored the frontier science of the microbiome and its relationship to human health, cognition, and behavior. His writing combines rigorous engagement with primary scientific literature, narrative craft, and a commitment to making complex science accessible to readers without specialized training.

The Electric Self grew from his conviction that the neuroscience of the past half-century is producing a revolution in human self-understanding as consequential as any in the history of science, and that the full implications of this revolution, for medicine, for technology, for law, for philosophy, and for the deeply personal question of what it means to be a conscious, embodied self, are not yet reaching the broad audience they deserve. He hopes this book contributes, however modestly, to that conversation.

Page | 288

www.ingramcontent.com/pod-product-compliance
Lightning Source LLC
LaVergne TN
LVHW022035200926
848481LV00003B/702